Whales In The

The use of metaphors in therapy

by Dr Jonathan Lloyd

Preface by Professor Michael Jacobs

Copyright © 2018 Jonathan Lloyd

ISBN: 978-0-244-41599-0

All rights reserved, including the right to reproduce this book, or portions thereof in any form. No part of this text may be reproduced, transmitted, downloaded, decompiled, reverse engineered, or stored, in any form or introduced into any information storage and retrieval system, in any form or by any means, whether electronic or mechanical without the express written permission of the author.

PublishNation
www.publishnation.co.uk

To Gill, Jasmine, Ellis, Theo and George.

I am like....

*I am like the wind
I am the train on the tracks
that runs and runs and runs*

*I am from you
me
us
them*

*I am like the glue that binds
the magnet that repels
the missing in the fog*

*I am like my child
I can play
I can smile
can cry
I can find strength from here*

*I am like a bird
I can fly
I can drift
as high as the cloud
I can burrow deep into the cloud*

*I am always there
ready
on the shelf
in the dream
in this moment*

I am like the creator
I can change
colour
shape
Your world and mine

I am like the gift
the chameleon pathway to your mind

I am like the knot in your gut
I rest in your heart
I rest in your neck
like the blade in your side
I hold the dreams you cannot tell

I am like
what you are like
change me you
our hills caves and dance floors

I am like the monster sleeping in the dark
that can lead to doors
doors hiding smiles behind

I am like the crack in the cult
the safe dungeon
the shiny hub
the frozen rose

I am like the tissues in the box
I am hope

I am like.

Preface
By Professor Michael Jacobs

Language is lethal. Okay, 'lethal' may be too strong an adjective to describe language, but because my adjective is a metaphor (at least as applied to the noun 'language'), it illustrates well my response to Jonathan Lloyd's richly argued and illustrated book. Not that his book is at all lethal – the opposite in fact. I will explain.

In some research into the effects of metaphors on communication, it has been suggested that the choice of metaphor influences the way people subsequently think. Researchers, Thibodeau & Boroditsky (2011) composed a short article about a fictitious city that had seen an increase in crime over recent years. The article began either with the phrase 'Crime is a beast ravaging the city of Addison'; or 'Crime is a virus ravaging the city of Addison'. Half the participants were given one version, half the other. Having read the rest of the article, which in both scripts was identical, the participants were asked what counter-measures should be taken to address the situation. Although overall both sets of participants favoured more punitive measures over educational and welfare programs, those who had the 'beast' version showed a stronger leaning towards stronger punitive measures than those who had the 'virus' version. The conclusion was that a single metaphor can influence thinking.

Of course the research findings have been challenged by Steen et al. (2014). For example, even though the rest of the article was identical in both cases, it was full of other metaphors. How did that metaphorical language influence the participants? Suppose the alternative beast/virus metaphors came at the end of an article rather than at the beginning – or what about comparing the use of a 'loaded' metaphor' with no metaphor at all?

Nevertheless, the conclusion of the original research ties up with this book on how influential metaphors are. It also reinforces the necessity of being attuned to what the particular significance of a client's metaphor imagery means for them, and how a therapist-introduced metaphor needs to be carefully chosen if it has not been part of the client's language. When I say that language is lethal, I mean that it can completely mislead. Misunderstand the client's metaphor, or introduce a metaphor that makes no sense to a client, and the misinterpretation of the metaphor by either participant can even be damaging. It can block progress or it can advance change. I take the point that Arlow (1979) makes, that metaphor can be a way of distancing oneself from powerful and potentially overwhelming affective experiences. But used well, that is the great virtue of metaphorical language, since metaphor acts as a transitional space, in Winnicott's (1953) terms, a safe way of introducing or describing difficult material, a distance that can gradually be reduced as the therapist sensitively works with it.

Jonathan appreciates this completely, showing how transformative metaphors can be. While I am not sure I agree with him that some clients do not like to use metaphors, since we all speak all the time in metaphorical terms, he understands how careful therapists have to be in the way they respond to the images clients use, and how far they introduce their own metaphors. And what he shows is just how much language, used well, can illuminate past and present, as well as the therapeutic process.

Michael Jacobs, Swanage, 2018

Contents

INTRODUCTION..1

CHAPTER ONE: THE CONSTRUCTION OF METAPHORS IN THERAPY..11

CHAPTER TWO: METAPHORS OF MOVEMENT..........28

CHAPTER THREE: METAPHORS OF ILLUMINATION ..60

CHAPTER FOUR: METAPHORS OF HOPE..................68

CHAPTER FIVE: METAPHORS OF RELATIONSHIP......80

CHAPTER SIX: MAJOR MODELS OF THERAPY & METAPHORS..93

CHAPTER SEVEN: BRINGING IT ALL TOGETHER, PRACTICAL ASPECTS 113

CHAPTER EIGHT: NEGATIVES AND POTENTIAL PITFALLS..124

CHAPTER NINE: CONCLUSION................................131

GLOSSARY..135

REFERENCES..137

INDEX...155

INTRODUCTION

I firmly believe that the essence of therapy is relationship. Without relationship there is no therapy, no connection, and no mirror to view oneself. As Michael Jacobs highlights wonderfully, language is an essential component of relationship and metaphors form an important part of our language. It feels to me that to be human is to be metaphoric, to want to connect and make sense of yourself, the other and the world.

When we speak and when we write we all use metaphors: 'The team was hungry for goals', 'The elephant in the room'. A metaphor is the substitution of the meaning of one thing for another. And then there are similes, which resemble metaphors but use the word like: 'He was like a bull in a china shop'. These structures have been used for millennia; indeed we have our metaphor for 'first-ness' in the term 'Big Bang'. The metaphor has saturated our language throughout history, from medieval allegories to Shakespeare. The Renaissance poets often linked natural objects and words; as Alan of Lille penned: 'All of the world's creatures are like a book' (Cooper, 1986). Wordsworth in Wordsworth & Coleridge (1800) proclaimed that poetry with its metaphors is the 'first and last of all knowledge'. Modern literature, poetry and music are flooded with metaphors. We continue to express ourselves, reason and think with metaphors; for example, the 'Worldwide Web' with its 'links', 'pop-ups' and 'drop-down' menus. It appears that the more complex our world becomes, the more we embody our ideas and concepts in metaphors (Geary, 2011).

Metaphor is vital to human communication and thus imperative to the world of therapy. For example, from a Human Givens perspective, Griffin & Tyrrell (2013, p.24) suggest that: "the brain is a metaphorical pattern-matching organ … we use pattern-matching and metaphor to communicate with others and to build our understanding of the world… Metaphorical communication is an intrinsic part of the way all human beings understand and communicate experience." Griffin and Tyrrell go on to claim that instinctive templates for behaviour for humans to survive in such diverse environments can only be identified in an approximate way, and metaphors provide an essential method of linking one pattern to another. To achieve our intelligence and survival, instinctive patterns learned from an early age (even before birth) need to be sufficiently flexible (metaphoric) to allow for a wide range of environmental variation.

My interest in the use of metaphor in therapy has evolved over many years of working both as a counsellor and hypnotherapist. It was during my hypnotherapy training course, perhaps surprisingly, that I was introduced to many psychodynamic concepts such as subconscious process, secondary gain, resistance, the importance of childhood and metaphor. Linked to the idea of metaphor usage in therapy was David Grove's utilisation of 'Clean Language' (Grove & Panzer, 1989). David Grove was a counselling psychologist who was popular with both hypnotherapists and counsellors as he had promoted maturing clients embodied metaphors to help deal with trauma. He had worked with many Vietnam War veterans in the United States.

As a busy practitioner engaging in both counselling and hypnotherapy, I found myself working with metaphor in both forms of therapy. Kincheloe (2001, p.687) suggests that "the frontiers of knowledge work best in the liminal zones where disciplines collide." I am fascinated by the 'common ground' that metaphors occupy.

In 2009 I was in a dilemma. I was commencing a Professional Doctorate in Counselling at the University of Manchester. I felt that I had something interesting to say about the liminal space between counselling and hypnotherapy. When using a metaphor, for example, clients appeared to enter an altered (focused) state and engaged in a creative way to resolve their issues. However, there was a language barrier between the two models. Many counsellors, including some lecturers, didn't like the idea of altered states in the counselling room. "This can't be proper counselling if your client drifts off." This world-view was directly in contrast with my own experience of working; in fact, both my clients and I reported that these moments could be cathartic and useful. How could I bridge these two similar, but often very different models? The answer soon came to me: metaphor!

I noticed that hypnotherapists and psychodynamically-oriented therapists appeared to use metaphor extensively. On this basis, I was able to convince my academic supervisor and the research panel that a heuristic study (Moustakas, 1990) of the use of metaphor in counselling and psychotherapy was viable and valid research.

This book reflects my experience as a researcher and a practitioner. In parts, it is practical and includes many case studies (names and details have been changed to protect clients' confidentiality). Also, you will find a more academic aspect to the book, which incorporates findings from my own and others' research. I hope that you, the reader, gain some insight on this fascinating topic. Please do not receive any part of this book as complete truth. I am merely reporting on what I have found useful (and also on what was not helpful) along with the thoughts of others on the subject.

In my research I was interested in hearing from experienced counsellors and psychotherapists who either used metaphors extensively or believed that they didn't use them at all. I found several of the former and one of the latter. Interestingly, after the interview with the latter he realised that he used them extensively with clients; this only became apparent to him during the research interview. My first finding was that metaphors are pervasive in counselling and psychotherapy, as they are in everyday language (Lakoff & Johnson, 1980).

Throughout the book I refer to 'my research'. This reference is to identify the source rather than express any type of ownership. I acknowledge that the research I carried out belongs to many people including the research participants, my academic supervisors, my fellow students, my clients and my family.

As a researcher who is also very much in practice, my experience of working positively with metaphors in therapy is reflected in this book. Clients 'take journeys down rivers with the spirit of their mother cleaning and healing the water before them'; they 'meet their authentic selves in meadows and play with the rabbits'; they 'talk to the devil behind them whilst the angel looks on'; they 'blacken out faces of the girls at school who watched as they fainted at the age of fourteen', and they 'face whales in the desert to get to the exciting town beyond'.

I have experienced a varied spectrum of metaphors working therapeutically for clients, and the outcome has appeared advantageous in the majority of cases. However, occasionally nothing happens. I am also aware that metaphors, for some, can have adverse results and I will review the possible reasons for this in a later chapter.

The above image was brought into my therapy room by a client who was suffering from severe anxiety. He had created this metaphorical representation to show me how he was affected by his 'inner voices'. I believe that this opened a dialogue about how he wants to be seen as the 'everyman' James Stewart. Stewart was an actor known for his portrayal of the average American middle-class man (Munn, 2013). I believe this representation led to a meaningful and fruitful exploration. We were both able to access from a shared cultural perspective the metaphor of 'hear no evil, see no evil, do no evil', and this became part of our narrative throughout the therapeutic process. His metaphor provided rich material for our discussions, far more than simply saying 'I hear these negative voices'; it improved our relationship in that I knew more about him and it shed light on his inner world. This was a metaphor of relating and illumination. Humour was also introduced to a previously bleak topic. The metaphor seemed to become a catalyst for more creativity, which appeared to help with his self-esteem and also externalised what was previously internal.

I recognise that I am a white, middle-class, British, middle-aged man working as a therapist and, as such, the metaphors that I introduce in this book provide information about myself and my cultural identity. One of the reasons I propose that 'the three monkeys metaphor' appeared to work well was that both my client and myself understood the metaphor culturally in a similar way. It may not have worked so well (or at all) if my client had not understood the metaphor, or their culture held a different meaning for it. I know that my metaphors are influenced by the time and place from which I come and the people that I meet. This way of considering metaphors fits with a social constructionist epistemology, which is grounded in the understanding that it is individuals and groups that participate in the construction of their perceived social reality (Burr, 2003). I believe that metaphors are culturally and relationally specific (Lakoff & Johnson, 1980).

I recently attended a mental health in education workshop in which an eminent historian introduced the metaphor of a waterfall.

He said that we are letting far too many of our children crash to the bottom of the waterfall where therapists are doing a good job but too late. We should be working at the top, preventing their fall. This metaphor appeared to work on some levels.

- The visual image gained access to our unconscious processing or the 'right brain'.
- The metaphor bypassed the defences of his audience. He did not directly accuse the audience (mostly teachers) of doing a poor job. He fostered a relationship with the audience as he could illuminate what was happening in the education system in the way they perhaps had not thought of beforehand.
- The waterfall offers the potential for movement. You can imagine helping those at the top before they

encounter the long and dangerous drop. I even imagine building a dam at the top!
- With the potential for movement, there is hope.

The advantages of metaphorical language are also apparent in a therapeutic setting. On a meta level, the use of metaphors in therapy represents the main components of what we are endeavouring to achieve most of the time as therapists: relationship, illumination, hope, and movement. Thus, I propose that there are:-

- Metaphors of relationship (conversational metaphors which reveal empathy in communication and improve rapport).
- Metaphors of illumination in which the client discovers more about themselves (where subconscious material comes into conscious awareness).
- Metaphors of hope (where hope is engaged with gently in the therapeutic relating in a non-direct way).
- Metaphors of movement (where the metaphors are developed and matured within the therapy that can engender a change in the client's thought process, intra- and inter-personal relationships, emotional connections and behaviour).

I understand that this categorisation of metaphors in therapy is just one way to consider this topic. Each type of metaphor is intrinsically interlinked. For example, the relationship in therapy is critical and an essential base to build upon; without an appropriate relationship, therapy cannot be conducted effectively. Metaphors of relationship and illumination can be developed into metaphors of movement or hope.

This concept, as it relates to counselling and psychotherapy, is intertwined with the rest of this book and forms its basis and structure. The focus is on how therapists operating under a wide variance of theoretical models approach metaphors that arise within the therapeutic encounter.

I suggest that the use of metaphor in therapy offers an alternative, less threatening, symbolic and non-direct way of revealing unconscious material. Dwairy (2009) proposes that metaphors create a therapeutic environment that allows the client to make alterations in their belief system while remaining within their family, culture, and religion. The indirect nature of metaphor invites the possibility for change that is perceived to be safe and non-threatening to the sense-of-self-in-the-world, although I would add this is only the case if the metaphor is mutually understood. If metaphors 'hit the target' they improve communication, understanding and rapport; if they miss then this isn't necessarily a negative. Metaphor and culture have a fundamental connection as the communication of shared meaning always carry cultural overtones.

In summary, this book aims (by considering metaphors in language) to improve therapeutic communication and relationships and not to act as an additional 'technique' or tool for your toolbox. Alan, a participant in my research who originally insisted that he didn't use metaphors with his clients, succinctly explains:

"For me, it's always been a case of communication. I try to demystify things, I try to use pretty straightforward, simple language and avoid a sort of technical psychotherapy-type word. I use metaphor to describe what I mean or to paint

pictures. It's a bit like colouring stuff in. If you exchange words with a client and you can come up with like a black-and-white picture, bringing in some metaphor can colour it in. It helps clients to remember stuff as well – when I use metaphor they remember things and they take things on board more effectively. And I know that because they'll then repeat some of the metaphors in subsequent sessions and say that they do find it useful to think about things in that particular way... For me, yes, it is about immediacy and appropriateness."

All the poems and images in this book have been created by the author (except the three monkeys image with permission of his client).

CHAPTER ONE

THE CONSTRUCTION OF METAPHORS IN THERAPY

In simple terms, the metaphors introduced into a therapy session must either be presented by the client or the therapist. I think it is a little more subtle than that, and I will discuss this shortly. Let us focus initially on client-generated and then therapist-generated metaphors.

Client-generated Metaphors

Meier & Boivin (2011, p.64) describe client-generated metaphors as: "idiosyncratic and very personal and transformed within the world view of the client". When working, I am very interested in my client's metaphors. They have the potential to be 'gold dust', a precious gift from the client, emanating from their language, culture and world-view, an unconscious attempt by the client to say: "This is how it is for me right now."

Client-generated metaphors represent the client's internal construct, which contains meaning and significance for them. Clients will often introduce a metaphor into the dialogue: "I am at a crossroads"; "He steamed into the house like a locomotive"(Kopp, 1995); or "I am stuck in the mud." All of these metaphors are an opportunity to investigate the client's world. Ricoeur (1986) recognises that to be advantageous the metaphor needs to be isomorphic to the problem, the story and the situation of the individual himself so that he can recognise himself in it and find new ways to perceive his difficulties.

Richard Kopp (1995) is seen as a principal author in the field of the use of client-generated metaphors in counselling and psychotherapy and has undoubtedly influenced the way in which I work. Kopp, a clinical psychologist, and Adlerian psychotherapist, highlights the various forms of metaphors used in therapy and identifies two broad categories: client-generated metaphors and therapist-generated metaphors. The Adlerian notion influenced him to believe that one's lifestyle is representative of imaginal and metaphoric cognitive modalities (Adler, 1927).

Kopp (1995) sets out a staged approach to his metaphor therapy:

- Noticing clients' expressed metaphors.
- Exploring the visual images relating to the metaphor.
- Expanding this to sensory images and related feelings.
- Inviting a change to the metaphor.
- Linking the metaphoric change to the real-life issue.

In his model, Kopp listens for the metaphors that a client relates to their life then merely suggests that he or she can change the metaphor to get what he or she wants. I would counter that this could be regarded as a reductionist and simplified view and does not honour the client's right to remain within their original metaphor. Kopp also promotes a method of linking bodily sensations to early memory and developing the metaphor of the memory, which can have connections to the client's current issues. He acknowledges that following a linear model can be frustrating for experienced therapists and is somewhat more prescriptive as a process than the Aeolian mode (Cox & Theilgaard, 1987) or indeed the ways that I offer in this book.

Research supports the positive aspects of focussing on the client's metaphors. In a Canadian study (Mathieson & Hodgkins, 2005) that used a narrative methodology to explore identity development in adolescent girls, ten teenage girls, aged thirteen to eighteen, who were in a hospital eating disorders treatment programme were interviewed. Mathieson and Hodgkins suggest that the focus on the client's metaphor assists the counsellor in building a relationship with the client, helps the client to give new meaning to their problems or experiences, and provides the client with new possibilities or solutions to their problems.

The teenage participants offered rich metaphors to describe how changes had occurred. They discussed the parts of themselves that wanted to change, using the 'self-as-a-container' metaphor. Various forms of rock-bottom were revealed: physical rock-bottoms when weight reaches an intolerable level, and emotional rock-bottoms. One participant identified her eating disorder as a separate person inside of her that shrank during recovery. Metaphors common in the field of addiction, such as 'finding myself', 'journeys' and 'roads to recovery', were identified. The authors of this paper recommend that counsellors listen to their clients' metaphors for change and consider the client's agency within them, as even what initially appear insignificant metaphors can be powerful vehicles for change. They also warn that counsellors should not force their world-view or their metaphors on their clients, although metaphors offer a potentially useful bridge between the client and therapist (Wickman et al., 1999).

This is an impressive piece of research, bearing in mind that clients with eating disorders often have concretised metaphors of self, which can be difficult for the client to change (Skårderud, 2007). The concept of concretised metaphors may help us to realise why anorexia can be difficult to understand; the client may be difficult to engage with because she or he is trapped in the concreteness of body symbolism. I will discuss concretised and sticky metaphors later in Chapter Three, Metaphors of Illumination.

From the Cognitive Behavioural Therapy (CBT) 'camp', Stott et al. (2010) agree that there are good reasons to pay close attention to client-generated metaphors as they have the client's attention and the client is attempting to understand their world and problems abstractly. Also, the metaphor may reveal relevant information about the client to both the client and the therapist, and it is safer to talk about issues metaphorically (Freud, 1917; Cox & Theilgaard, 1987). In my experience, this can enable the metaphor to continue and develop over many sessions, if not throughout the whole therapeutic process. Naziry et al. (2010) discovered a marked improvement for clients reporting depression when their metaphors were engaged with by the therapist in a study focusing on CBT.

From my research, the participant Anna alludes to the improved sense of self that can result when a client coins their metaphor:

"So my fantasy is when a client comes up with a metaphor they've had to struggle through something to express themselves. To get their thoughts and their verbalisations around something that's intangible and going on for them is in itself is incredibly healing and potent for them. All those neural pathways have been firing and connecting."

There appears to be a good reason to create an environment in which clients feel comfortable talking about their lives metaphorically and to help them to investigate and develop their metaphors.

Therapist-generated Metaphors

Sometimes I feel an intuitive urge to offer my metaphor to help the therapeutic dialogue or provide an alternative way of seeing the client's issues. Some therapists have a bank of metaphors for particular presenting problems or situations that present themselves. CBT literature promotes this method of revealing metaphors to clients (see Stott et al., 2010).

I tend to introduce my metaphors with an element of caution, although I have found that clients report that they have been useful for them in particular contexts. That is a benefit of metaphors: they can be introduced to clients in a non-invasive way. Clients can pick them up and run with them if they connect with their world-view and experience, or they can be left without any sense of threat. Examples of generic metaphors I might use include 'the wheel of relationship' or 'the well-being tree'.

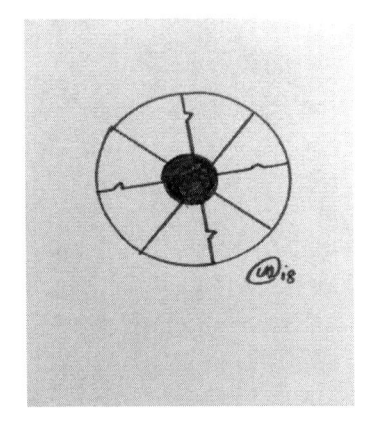

The wheel of relationship is a metaphor that I learned during couple-counselling training. When working with a couple (or an individual) who are encountering relationship difficulties, I will talk about a wheel that has spokes and a hub. The spokes represent all the aspects of the relationship: money; children; friends; work; in-laws; health, etc. Sometimes the spokes get bent, causing the wheel to buckle and wobble. The hub at the centre of the wheel represents the love, care, time, sharing, texts, open communication, intimacy, sex, etc. between the couple (or, as one client called it, 'the shiny bit'). The larger the hub, the easier it is for the wheel to tolerate bent spokes. A small or non-existent hub cannot cope with many bent spokes at all. Couples seem to grasp this idea and often report that they have been 'hub-building' during the previous week.

The well-being tree is used with clients with a strong archetype of carer, who always put others first, often at the price of their self-care. I will also use this metaphor in well-being workshops, working with head-teachers or carers of people with dementia. I will draw a tree; the roots are the part of the system that needs nurturing, watering and feeding. If you look after the roots (self), the tree grows a sturdy trunk and branches (the teaching staff). A happy by-product of healthy roots and branches is the fruit or leaves that flourish and grow. When focusing on the leaves (children/cared for) at the expense of the roots and branches, there is a chance that the whole tree will not survive. A similar metaphor is the message on the aeroplane to put on the oxygen mask yourself before you help others. This metaphor may not be an appropriate one to use in all situations (with narcissistic clients, for example) but it seems to resonate with most clients.

Cox & Theilgaard (1987) are regarded as critical theorists concerning the use of metaphor in psychotherapy. Unlike other theorists who promote the universal response by the therapist to their client's metaphors (Grove & Panzer, 1989; Kopp, 1995), they suggest that the therapist answers an 'irresistible call' to respond to their client's disclosure of experience. According to Cox & Theilgaard (1987), sometimes the language of the client is unambiguous although the therapist is also encouraged to respond to the affective loading of the client's narrative or even to their silences.

The impact of the inner-world of the client on that of the therapist is the catalyst for movement into the Aeolian mode, a sharing of poetic language, metaphors and images. It is therefore difficult to identify where the metaphor begins in this mode of working – with the client or therapist. I would suggest that they appear to be born in the intimate therapeutic space. Cox & Theilgaard (1987, p.43) confirm the relational and mutuality aspects of this way of working: "although creativity is involved, during the actual unfolding of a therapeutic session, the therapist is not aware of 'inventing images'. On the contrary, he perceives them." There is also an acknowledgment that if the therapist's metaphor is aptly timed, the client never regards the image as intrusive or inappropriate.

When working in this mode, the therapist is always working with perceived new imagery rather than waiting for their client to verbalise a metaphor. Brian, a participant in my research, highlights a possible difference in that psychodynamically-informed psychotherapists are more likely to use their own metaphors, unlike more person-centred counsellors who may wait for their clients' metaphors: "*I am more likely to create the metaphor and visualisation as we go along. I personally feel that they create a level of safety that may or may not be there if somebody just creates their own images.*"

Tom Strong (1989) suggests therapists require particular skills to utilise client- and therapist-generated metaphors as a vehicle for change, including tuning in to their client's language and engaging in a dynamic, interactive process.

Strong's model (1989) is different to many other models of counselling as he sets out three types of therapeutic interventions and processes that explicitly use metaphors.

- Explicating what is implicit: this is a process of reflective listening. The client's metaphors are explored and matured, theoretically resulting in increased self-disclosure.
- Therapeutically extending or modifying the client's metaphor: this is where the counsellor works directly with the metaphor generated by the client, usually through the form of guided imagery.

- Creating and delivering therapeutic metaphors: this intervention involves the delivery of the metaphor by the therapist. By listening carefully to the client the therapist, through metaphor, offers an alternative context or container for the client's 'problem', which should be in tune with the client's language. The metaphor therefore offers a 'recalibration' of the problem and the client starts to see their issues differently. At this level, the counsellor avoids using directive language, similar to Clean Language (Grove & Panzer, 1989); theoretically this in turn promotes internal processing by the client to construct personally appropriate meanings in response to the counsellor's ambiguous words.

In a UK-based qualitative study, Bayne & Thompson (2000) set out to review the use of metaphors in counselling through the lens of Strong's model. This study focused on two questions: are the three strategies used by experienced counsellors, and are other approaches used? They found that some counsellor responses did not fit within the specified categories; often the client's metaphor was remembered for later use. Seven experienced integrative counsellors, working in various settings and without prior knowledge of Strong's model were interviewed and, while Strong's model was utilised, the therapist-introduced metaphor was identified only once in thirty-five examples, and even this was informed by a 'tunnel metaphor' introduced earlier by the client. The authors hypothesised, without any evidence, that the absence of therapist-generated metaphors was due to the lack of psychodynamically-trained counsellors within the sample. In my research experience, psychodynamically-oriented therapists are more likely to offer their metaphors to their clients.

If you offer your own metaphors, following Strong's (1989) model, using as much of the client's language as possible is important. I would also include being mindful of the client's culture. The cultural influence on metaphors is highlighted in the influential work of Lakoff & Johnson (1980). They argue that different cultures give rise to mixed metaphors due in part to the fact that they are associated with different human experiences, and they provide various linguistic tools to characterise things. Lakoff and Johnson also argue that cultures produce conceptual metaphors that form our way of seeing the world. The very nature of the conventional metaphor is that we are not ordinarily aware of its figurativeness; the ways it affects meaning are therefore subliminal. Many theorists agree that not all metaphors are universally understood and there are cultural differences in many metaphorical expressions.

Kövecses (2002) suggests that inferences are unavoidable when either interpreting an English metaphor such as 'Manchester City slaughtered Manchester United' or an Inuktitut metaphor, 'inuktaiwa', which can either mean 'killed him' or 'took him as servant'.

Yu (1999) identifies that diverse cultures relate to the time-is-space metaphor. Interestingly there is a difference between some Western cultures that frame the future spatially as being in front and other Eastern cultures that view it as being behind. I remain sceptical that therapists and clients understand all of each other's metaphors, particularly when they originate from a different culture (Holland & Quinn, 1987). I recall one client from Africa who 'googled' the metaphors that I had used after each session to gain a better understanding of them.

It appears to me that therapist-generated metaphors can be extremely useful if they are apt, well timed and can be mutually understood by both parties. There is a sense that both parties can develop an understanding of enhanced self-awareness through this way of working. While Kopp (1995) primarily promotes the development of client-generated metaphors, he does not necessarily have a problem with therapist-generated metaphors as long as they resonate isomorphically with the client. For example, McCurry & Hayes (1992) suggest that therapist-introduced metaphors are most effective when they are understood by the client and are drawn from the 'common sense' world of everyday objects.

The understanding of a metaphor is dependent on the familiarity with both the topic and the vehicle of the metaphor and the 'distance' between the two. McCurry and Hayes provide a therapeutic example of working with a child who was moved into new foster home, the third in as many years. The therapist introduced a simple garden metaphor in which a plant is dug up from a neighbour's garden and replanted several times; with care, water and sunlight the plant grows and flourishes. The therapist gave the child cuttings from her garden to allow the child to experience replanting new cuttings that would enable them to grow. This beautiful example of a therapist-introduced metaphor was mutually understood and developed (even in a practical sense) by the client and therapist and offered an opportunity for reframing the situation into a more hopeful one. (I note that the therapist took a risk that the plant remained healthy.)

Pitts et al. (1982) trained therapists to use metaphors in therapy and they found that the therapists saw the construction of metaphors as emotionally gratifying (it was also suitable for the therapist to adopt a client's metaphor). While this study did not research the views of the clients, the focus on the *construction* of the metaphor by the therapist rather than the *outcome* was viewed as beneficial.

Based on many years of working as a psychoanalyst, Sapountzis (2004) found that appropriate therapist-generated metaphors were extremely useful when working with children who were withdrawn and disengaged. The common theme here is of being mindful of the client, their language and culture if you want a positive outcome following the introduction of your metaphor. Also, the non-direct aspect of metaphors enables the client to more easily reject your metaphor or amend it if it does not fit for them. As Stanley-Muchow (1985, p.47) helpfully reminds us: "a metaphor must prove itself fertile in the mind". This suggests that it is a common understanding and mutual development of the metaphor that are the essential factors, whoever is the originator.

Co-constructed Metaphors

Alternatively, I suggest that the metaphors arise out of a shared social pool of symbols and meanings (Burr, 2003; McLeod, 2004) and are transacted within a co-constructed process and interactive relationship (Tebbutt, 2014). It appears to me that all metaphors are co-constructed between the therapist and their client in some way. Often, I will form a visual image in my mind that resonates metaphorically with the client's world; this is offered to the client who then has the choice of whether or not to run with it. Therapists also reported this during my research.

Füredi (2004) argues that psychotherapy itself appears to have a culture of its own and the metaphors and language of therapy appear to have leaked into Western mainstream culture. I see this as more of a two-way process, where metaphors are imported into the culture of therapy, often mature within that culture and are then exported to the culture of non-therapy. This is highlighted in the study by Angus & Mio (2011), which indicates that healing metaphors co-constructed and developed by counsellors and clients in a hospital setting were subsequently also taken outside the relationship and used with medical staff.

In my research findings, many participants commented on the construction and development of metaphors in their work. By way of example, Maddie said: "*When somebody is talking to me, I have pictures going on in my mind, I have images flashing across my brain. So I use images and weave them into, it's hard to think of an example. Say I am talking about somebody feeling tumultuous I start to talk about tornadoes and sort of weave that into something. That's the most exciting thing, when they realise that you've understood it, and you can translate it into ways that they can get.*"

Jane said: *"If they start speaking in metaphors and I pick it up, or if it's something that they're talking about and a metaphor springs to my mind, I offer it and then see what they do with it. If they look at me as if I'm barking mad, then I'll back off a bit and perhaps try again a little later and perhaps modify what I've done. But if they pick up on it, you know, they say, 'Yes, that's right', I go with it more."*

There appears to be a common theme of therapists sharing visual images with their clients to build a sense of understanding and potential movement. The metaphors can be developed by both parties once the shared understanding has been co-created. What would be interesting would be to find out if clients are aware of offering visual images, and maybe highlighting this concept during the therapeutic process.

I will use a practical allegorical example to explain my idea of co-created metaphors. On Boxing Day my wife and I made a hotpot (a cross-cultural term used to describe a particular meal). To begin with, my wife prepared the vegetables and part boiled them. On my return from a football match, I added the vegetarian mince, seasoned it and continued to cook it. I then baked some crusty bread. We both added our red cabbage, more salt and pepper and enjoyed the meal together. The point that I am raising here is that Gill started the process (in which I was later involved) and she claimed significant ownership by having the idea of the meal and bringing in the ingredients, although effectively we made the meal together. The construction of the meal was mutual and emerged from a shared understanding of how hotpots are created and that we like them. You could also argue that this shared understanding is the result of our being in a relationship. In the same way, whether the therapist or the client introduces the metaphor is not as important as the shared meaning and mutual development. I am proposing that co-constructed metaphors are born and developed in the relationship. Gill and I know that we both like vegetarian food, and that vegetables would be a healthy option after Christmas Day's excesses and that warm food would be well received after a cold day outside. There was a mutual understanding before the meal was prepared based on our relationship of some thirty years and a resultant mutual understanding. I propose that the hotpot, along with the metaphors that arise and are developed in the therapy room, are socially constructed (Crotty, 1999). As Dalal

(1993, p.407) succinctly claims: "The social is present from the beginning". Empathy is a step-by-step process of relational co-construction (McLeod, 2004).

Through an existential lens, the term co-construction is described as: "...the reflection of being and object of reflection are defined through each other, they are co-constituted. We are actively involved in any experience and what we experience is co-constructed by us and by the object/person that we encounter – any experience of relationship says as much about me as it does the other, it is a co-constructed relationship" (Van Deurzen & Young, 2009, p.208). Thus, existentially relationships are always in the process of co-construction. If this forms an element of truth, then maybe metaphors used between parties reflect this and become co-constructed too?

People take part in 'language games' in conversation (Wittgenstein, 1953; Hobson, 1985). Wittgenstein (1953) talks about speech as 'fight and play' (playing with metaphors?) or speech as creating 'social bonds' (the co-creation of metaphors?). Neimeyer (2002, p.51) argues from a psychodynamic viewpoint, that psychotherapy functions less as a context for self-analysis, rather "as a dialogical sphere for the co-construction of a new sense of identity within and beyond the therapeutic relationship."

Change in therapy can occur through the development of co-constructed narrative (White & Epston, 1990) and co-created metaphors form a significant part of this process (Goncalves et al., 2009). I would contend that co-creation of metaphors, in this context, also connects to the interpersonal, intrapersonal and transpersonal aspects of moments of a deep encounter (Tebbutt, 2014). Cox & Theilgaard (1987, p.41) quote the words of a client who had been working with their therapist using metaphors in the Aeolian mode. I believe that the stanza quoted powerfully highlights the co-constructive nature of working in this way: "not our things, not our 'bits and pieces', but our belonging – yours and mine – our 'belonging', to each other – our belongness – Together, Home."

The following image displays my reflections on the social and cultural aspects of metaphors in therapy. Here, in part, I refer to the inevitable nature of the relational aspect of the co-construction of metaphors and suggest that the client and therapist bring in their metaphors from their culture to the subculture of therapy. These metaphors are then co-constructed, mutually developed and have the potential to return to the outside of therapy culture. I would offer that the subculture of therapy has the potential to 'ingest' through symbolic interaction those cultural metaphors previously owned separately by the client and therapist.

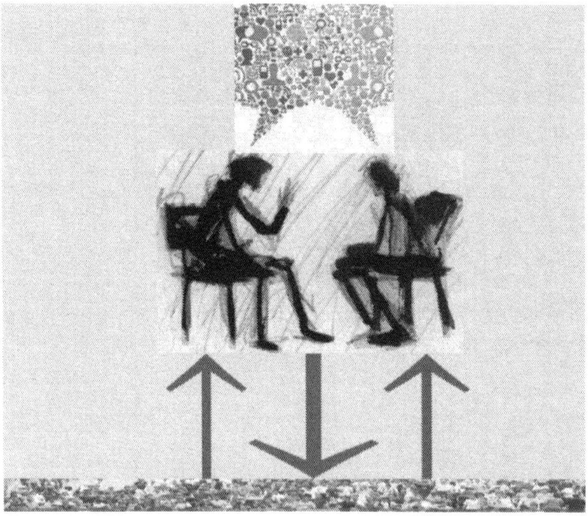

CHAPTER TWO

METAPHORS OF MOVEMENT

In this chapter, I investigate *metaphors of movement* in therapy and how they are developed. I will highlight the importance of the therapist's language and how the words they use can aid the progression and development of the metaphor.

In my experience, it is the development, movement and segue of the metaphor that holds the most potential to promote therapeutic change. Of course, sometimes metaphors can become stuck, but there is useful information and the possibility for personal growth in the stuck-ness. Metaphors of relationship, hope and illumination (while unavoidably connected to potential movement) are discussed in more detail in subsequent chapters.

Metaphors of movement are useful in that they allow the client to perceive aspects of their experience in an alternative way without being threatened and thus defensive. I would define metaphors of movement as: "Metaphors which are matured and developed to such an extent that the parties involved become aware of alternative ways of viewing a particular phenomenon." Some behavioural or relational change in the client will often follow this change of perception, although this is not always the case. These metaphors involve flow and directivity of metaphorical images shifting and changing, directed by the therapy participants – what Hobson (1985) would describe as a 'moving metaphor'.

As I alluded to earlier, the client and therapist seem to co-construct and mutually develop the metaphor during their relationship. Cox & Theilgaard (1987, p.29) suggest that, when working with metaphors of movement, co-construction and mutuality are vital: "It is the impact of the inner world of the patient on that of the therapist and vice versa which promotes movement". I would also add that it is the impact of each other's inner worlds that creates the metaphor in the first place.

To highlight origination and development of metaphors of movement let me introduce you to Meredith.

Case study – Meredith

Meredith, a woman in her early 60's, entered the therapy room, the light pouring through the blinded window. She had offered many metaphors in previous sessions to explain her life situation, and metaphorical language appeared to feature strongly in her communication. For example, two weeks before she had said she felt stuck in 'grey metal' and couldn't speak. She could only view some form of Utopia across the river (which wasn't accessible to her). Through discussion with me, she had developed the 'grey' to a dance floor where she enjoyed to dance but still viewed freedom and beauty on the other side of the river. She could speak to a fish in the river; I pondered whether this was me but decided not to follow this route as I wanted to stay with her language and allow her to guide me.

During the last session, she had described darkness inside her. When I gently explored what the darkness inside her was like, she explained it was like a "dark monster", a monster lurking in a cave. She wondered if it was a message from her dad that bad things happen. She sensed that the beast did not want to come out, and she certainly wasn't going into this cave herself! Today she had a picture in her hands; it was the monster and also a young girl with silver sparkles around her. The monster was angry, faced away from the little girl and had no legs.

A representative illustration from my notes

At this stage, I could have asked her about her week and if anything had encouraged the monster to exit the cave. I resisted this temptation as my intuition was to stick with the metaphor. I reflected on what I had noticed and asked her what needed to happen. She said that the little girl was scared and needed to be protected from the monster. The little girl wanted to help it to be happy, to throw it a ball. This revelation led to a conversation about how the little girl's role was to ensure that everyone is happy (even monsters).

The following session, she brought in a different picture. I noticed that the girl was older and had more glitter around her. The monster had lost its spikes, was less dark, and faced towards her. She said that it needed to find its voice. The stitches which she had previously drawn on the monster's mouth needed to be removed. I reached for an eraser on my desk and handed it to her. She immediately, but carefully, started to rub out the stitching. At that moment she reminded me that one of her childhood issues was that she struggled to express herself. She had no voice. Once the stitches had been removed, she opened up about her relationship with her husband and her father.

A representative illustration from my notes.

Meier & Boivin (2011) suggest that metaphors can be categorised as either client or therapist generated; I have offered an alternative social constructionist hypothesis that metaphors are co-constructed by both parties. In the above example, Meredith introduced the idea of 'grey' to me although it is not known how she came to this metaphor. Was it her metaphor, or did we co-create this in our dialogue? What we do know is that the metaphor was mutually developed towards something more descriptive and potentially resourceful therapeutically.

During this process, I was careful not to influence Meredith's metaphor and I utilised neutral language with questions like "what is that like?" or "what would 'grey' like to have happened?" I would often address the metaphor directly (the girl or the monster, for example) and use Clean Language (Grove & Panzer, 1989). I believe the use of Clean Language by the therapist is useful when working in this mode and I will develop this idea later in this chapter.

It is interesting to note that as the metaphor extends Meredith grows up from a small child through to perhaps a teenager who can discuss relationships in a more adult way and needs less protection (glitter). In my experience, this is often the case when working with metaphors (the subjects get older). You could also argue that I was represented as the fish, the talking fish that occupied the liminal space between her reality and her 'utopia' (freedom or preferred scenario).

There are many interpretations one can make from this brief snapshot of my work with Meredith. I would like to raise some important points here:

- My work with Meredith was long-term (over several years). While she used metaphors in her communication, this type of mutual development work did not start until some months into therapy after a healthy relationship had been formed.
- Meredith was very creative and enjoyed working in this way.

- The mutually-developed metaphors improved our communication and rapport.
- There was no apparent obvious remedy to her issues in the metaphors, only an increased self-awareness that in turn led to behavioural and relational changes. (Although there are other examples when significant change occurred with other clients in one session.) She realised she had options.
- The metaphors helped bypass her defended past; it was easier to talk about the monster than her traumatic experiences.
- The metaphors flowed from session to session (fortnightly in Meredith's case).
- There was a sense of movement and development in the metaphors. This sense of movement seemed useful and perhaps created a sense of hope for her and me.
- With my support, and after some more time, Meredith felt able to explore the metaphors. They led to an exciting place, with lots of doors leading to adventures and exciting ideas for her to explore. She realised that she has many doors to explore, which I believe were doors into her 'self'; it was a journey of self-discovery, which in turn led to changes in beliefs and behaviour.

I find that I often go on a metaphorical 'journey' with clients in therapy; they often start in a dangerous or undesirable place and eventually find a place of hope, safety or healing. The words of Hoffman (1967, p.69) echo for me here: "we journey from the narrow place through the perilous place to a safe place". I note that we create this mythical landscape to journey through together. Meredith could have constructed a bridge over the river to the 'Promised Land' if she had been ready for that change at that time. It seems that both parties appear to display and share a substantial understanding of the metaphor. I then experience a mutual development of visual images, or feel senses that could be likened to the issue being addressed in therapy.

In addition to my own experiences, there is also evidence in the literature that the mutual development of metaphors can, in the main, be beneficial (Hobson, 1985; Marlatt & Fromme, 1987; Cox & Theilgaard, 1987; Goncalves et al., 2009; Angus & Mio, 2011). In the psychodynamic literature, 'intersubjective construction' is encouraged (Ogden, 1997) and some of the CBT literature identifies the benefits of mutually developing co-constructed metaphors (Spandler et al., 2013).

In the person-centred model, the reference to metaphors of movement is more subtle and is based on the therapist's attitude influencing the relationship and, in turn, the client. However, in the dialogue and process forms of the model, co-constructed dialogue, mutuality and co-creative relating are identified as crucial aspects (Bohart & Tallman, 1999; Tudor & Worrall, 2006; Sanders, 2007).

I will now introduce some case studies on working with metaphors of movement. In Chapter Nine, I look at the practicalities of working in this way.

Case Study ~ Prancing Sheep and Curly Wurlys

Pam was in her mid-sixties and had just retired. She was looking towards a healthy retirement with her husband, and her daughter had produced a new grandson. All was rosy in Pam's garden! She then contracted tinnitus (a condition from which I also suffer). Nothing like this had affected her before, and she was reeling with anxiety. "Why me! This is awful! I can't cope," were questions she kept asking and statements she kept making. She had a lousy time getting appropriate help from the NHS. By the time she reached my therapy rooms, she was extremely anxious (unable to sit back in her chair) and reported a distinct lack of sleep.

She was delighted to hear that I had managed to come to terms my tinnitus, which gave her hope and I was able to reveal deep empathy towards her. Over the weeks, we tried to find a metaphor which had the potential to shift her relationship with her tinnitus and pulsations from unwanted to accepted. Pam was quite literal, and we struggled to find a metaphor that worked for her; she tended to literalise. She couldn't leave the noise outside or change it to balloon (even if this was a Manchester City balloon, which connected to fun and success). I offered her a blank canvas on which to work. Rather than trying to use her imagination, I gave her a blank A3 page and a pen and asked her to draw her tinnitus in her ears and the pulsations at the back of her head. She surprised me by drawing very quickly and spontaneously two squiggly balls (like balls of wool) to represent her tinnitus in each ear and four distinct lines for the pulsating at the back of the head. I thought, 'Great; this is a golden opportunity to start to develop a metaphor of movement.' Within the same session, she was able to change the original image to prancing sheep and 'Curly Wurlys' as in the following reconstructed drawing:

Again some notes may help here:

- Creating a blank screen is a useful way to start a metaphor; it is clean and unpolluted. My offer not to judge her drawing ability also proved useful for her to start the process.

- With some clients, particularly those with more literal thinking, actually drawing the image is preferable to imagining it.

- This altered metaphor helped Pam to relate differently to that which she could not change. Prancing sheep are preferred to squiggles of noise and Curly Wurlys (a Cadbury chocolate bar) are preferred to stark straight lines.

- She saw the humorous side of these changes and reported the next week feeling much better and sleeping well. She described seeing some sheep in the snow, which reminded of her own 'prancing sheep'.

- The metaphor of movement was only part of her therapy. I think the empathy and hope I had portrayed through my own experiences were conveyed as being very helpful for her. This was also a metaphor of hope, in that she believed she could cope with the tinnitus on a long-term basis and her anxiety would keep reducing over time.

- The metaphor of illumination here (her original drawing) wasn't enough for Pam; she needed and wanted to change her relationship to the issue. She was already overly self-aware of how she was relating to her tinnitus and pulsations.

- Humour was introduced to a previously sombre and anxiety-provoking subject.

Case Study ~ Swinging Chair

Reconstructed image from my notes

I will be referring to 'Emily' later in this chapter when discussing the merits of Clean Language. She had originally come to see me for weight loss but, as often is the case, this turned into therapy about her childhood and her relationships in the past and now. She particularly had an issue with her mother who, she felt, controlled her. Part of the control, even though Emily was approaching fifty, was to overfeed her whenever she visited her flat.

Emily desperately wanted to gain some control in her relationship with her mother. We had been working on this for many months in various forms. She thought of the idea of imagining a fence around herself whenever she visited her mother's flat. This metaphor worked to some degree, but she kept forgetting to build the fence each time. With a big smile, Emily conveyed that she come up with an alternative. She had imagined putting her mother in a wicker chair that hung from the ceiling. Emily had the only key to unlock and release her mother. She let her down occasionally but would pop her back when she felt controlled and overwhelmed.

Interestingly, with this new thought process, her mother started to give her normal-sized portions, the same as she gave to herself (the mother was a very slim woman.) This reduced Emily's anxiety in the therapy room as she reported a physical change while at her mother's flat.

I would describe this metaphor as a metaphor of movement as it offered Emily an alternative way of being with her mother.

- I felt this was an extremely powerful session and was the start of her recovery and the end of our work together.
- The metaphor work didn't stand on its own and was part of a long-term relational therapy, in which Emily was encouraged to find creative solutions for herself.

- This metaphor was amended by the client outside the therapy room. This often occurs: metaphors born in the therapeutic encounter are exported and imported back and forth from within and without. This is empowering for the client and appropriate that they find out what works best for them in their 'real world'.
- Once Emily had imagined taking control in this way a number of times, she didn't need the swinging chair each time. This was enough to give her confidence that she had some control. Taking control was very important for her. I would suggest that her eating issues were a reflection of her lack of control. Indeed in the session following the 'swinging chair' Emily said she had stopped eating when she felt full.

Other examples

One of the core features of the metaphor of movement is to "take us on the way a little" (Cox & Theilgaard, 1987, p.12); there is a directional aspect of this way of working. The therapist's and client's responses to the metaphor of movement influence what follows and so forth. This can be a process within an individual session or over a series of sessions.

An example can be found from my research. Anna a psychotherapist reported a simple but effective metaphor of movement: "One client yesterday, who was struggling with feelings of terrible rage, said it felt like his life was like a motorway with three lanes on it. One lane curves off and goes to an angry place and one road curves off and goes to a place of rage where he is not conscious of anything. Through counselling he now has a lay-by where he can pull over, calm down and carry on straight ahead without being diverted. It's a perfectly contained, personal to him, expression of what is going on for him". I propose that this also reveals the bespoke nature of metaphors of movement.

The metaphor of movement can be swift, profound and occasionally involve an age regression of the client. For example, Tom's use of metaphor in the therapy room was intense. This client addressed a moment of profound shaming at school (which he believed was the cause of his sexual problems as an adult), by gifting his imagined five-year-old self with his (adult's) tattoo. The little boy felt stronger, more grown up, more able to cope. This is an example of a metaphor of movement where Tom and I co-constructed a metaphor (in which I used Clean Language) that represented a memory that could be reframed safely through the eyes of the adult in one session. Tattoos, on reflection, are used only on adults in our culture. I will allude more to age regression in a following chapter.

Another client, described in the introductory section of this book as the man with the three monkeys and James Stewart image, visualised his problem (social phobia) as being like a whale in a desert. It was too messy to go straight through the whale, he stated that he needed a stealthy option of going around the whale to get to the town where he could have an adventure on the other side. We mutually developed the metaphorical landscape. His mother was pushing him along in a wheelchair as though it were a marathon, his father was a 'road-bump' slowing them down, and I was a 'direction sign' showing him the way around the whale.

This metaphor mirrored the process of the therapy and helped us both monitor his progress about how close he was to the whale; it also appeared to help reveal the amount of resistance he had to dealing with the issue (he wanted to go around the whale rather than deal with it). The metaphor also helped explain to him how over-helpful/unhelpful his parents (and I) were being. He was appalled by the image of himself being disempowered, and this might have added to the leverage for a change.

We were able to review the metaphor in subsequent sessions and deconstruct the whale by removing its blubber and skin (possibly his anger and hatred towards himself and others). I was mindful that I was conspiring with helping him to avoid the whale (the difficult emotional process of reconnecting with his peer group).

I am pleased to report that this client has now been through the whale and is enjoying the exciting town on the other side.

Jane's Case Study – Scooby Do

Here Jane, a research participant, describes the development of a metaphor of movement with a young client.

This came about whilst working with a seventeen-year-old client with an ED (eating disorder). After building the relationship for a few weeks, featuring usage of humour, the client mentioned her 'dog'. She described a Rottweiler that would wait outside her bedroom door at night or outside the bathroom door when she was showering (sometimes climbing in the shower with her). Naturally this was extremely distressing and meant she felt trapped inside the rooms. She hadn't previously disclosed this dog to anyone else.

Aside from exploring who or what this dog represented, I decided the first step would be to 'move' the dog so she could at least pay a call of nature in the middle of the night, and also manage the stress she was experiencing when encountering it. The client agreed with this strategy and we spent a session 'ridiculing' the dog, thereby reducing the fear and power it held. We named it 'Scooby', visualised it with a pink bow round its neck, wearing a 'designer' dog coat and wellies, gave it a 'Gnasher' cartoon grin and made it 'run off' into the distance to the tune of Benny Hill.

Our rapport in the session was collaborative and fun – much giggling – but it served as a resource for her with the process of dressing the dog and removing it when she encountered it. Most of the time it worked for her. It also provided us with a private, shared discourse that became a running feature of our sessions... "Is Scooby wearing Burberry this week?" I believe this approach provided us with a vehicle with which to slowly unpick the meaning behind the dog.

Let us consider some of the detail of this enlightening therapy:

- In this interesting case study, I am struck by how the humour in the metaphor acts as vehicle for change. The 'energy' of a metaphor is similar to the energy of humour; it can change the therapeutic charge in the room.
- The client introduced the metaphor (a novel metaphor not mentioned beforehand to anyone) but this was based on a maturing therapeutic relationship and dialogue.
- Because of the depth of the relationship, the therapist was able to offer change to the metaphor and to move the dog.
- The ongoing metaphor gives a frame of reference for both parties to share a changing image that belongs to them. The client can start to take some control.
- The relationship is enhanced by the joint development of this metaphor.

Clean Language

The language of the therapist is essential when both he/she and the client are developing a metaphor, particularly a metaphor of movement.

Clean Language is a way of using words and forming questions in a neutral way that helps the recipient stay with their language and metaphors. It is a way in which, in the context of therapy and metaphors, the therapist does not influence the expression, stance and flow of the other. In regular conversations the use of Clean Language may appear a little strange but it feels natural when used when the other person is in engaging in their own story, narrative or metaphor. The therapist uses Clean Language to help the client move the metaphor. Questions such as: "What happens next?" or "What needs to happen?" are used; these are 'Clean' and 'moving forward' questions.

Many models of psychotherapy and counselling have paid different degrees of attention to the words used by therapists. The person-centred approach, developed by Carl Rogers, promoted a non-directive engagement with clients. Rogers believed that humans have an innate tendency to find their fulfilment. In Rogers (1980, p.117) he states that: "'Individuals have within themselves vast resources for self-understanding and for altering their self-concepts, basic attitudes, and self-directed behaviour; these resources can be tapped if a definable climate of facilitative psychological attitudes can be provided". To use a metaphor, Roger's Person Centred model was the spring from which the river of Clean Language flows.

I have already referred to the late New Zealand counselling psychologist, David Grove, who pioneered the use of embodied metaphor and Clean Language in counselling and psychotherapy in the 1980s, particularly for working with clients who have suffered a trauma. David Grove worked with some Vietnam War veterans in the United States. I understand that Grove's tutor was Ernest Rossi (formerly a student of Carl Rogers). I find that when engaging with metaphors of movement in therapy, the use of Clean Language is helpful in helping the client stay within the language of the trope. Further, this method of using neutral language and repeating the words of the client in the tone of the client can aid attuning towards a client, particularly those in an age-regressed state. For example, if the client accesses a younger self in the metaphor work and softens their voice, the therapist will also soften their sound when using Clean Language. This way of working will help the client remain at this developmental age throughout the metaphor development (Erskine, 2015).

I do find similarities in the process of Clean Language and David Grove's embodied metaphor-maturation model to the 'Focusing' approach detailed in Gendlin (2003). Gendlin regarded his work as being Rogerian, although some purists have commented that his work is too directive.

Gendlin's six-stage process is founded on the premise that "only the body knows your problems and where their cruxes lie" (p.11). The stages of Gendlin's model are: clearing a space; felt sense; handle; resonating; asking, and receiving. Once a 'felt sense' has been identified within one's body, then a 'handle' is identified which links with the felt sense.

These stages can be a word or phrase, but also can be an image, a metaphor or symbol as outlined in Grove & Panzer (1989). Following a period of resonating the connection between the felt sense and the handle, the final element detailed in Gendlin (2003) is a transaction of asking and receiving information from the 'handle', the embodied word or image. Again, this fits with Grove & Panzer (1989) where the metaphor is communicated with directly.

Gendlin (2003) does not refer to Clean Language although he does highlight the importance of the therapist not analysing, being mostly silent and "avoid(ing) forcing words into the felt sense" (p.55). Also, the "triggering questions" (p.104) have a neutral element to them, which aligns Gendlin's model to

Grove's. For example: "What is at the centre of it?" or "What is it doing?" (Gendlin, 2003, p.104).

Other models such as Neuro Linguistic Programming (NLP) have also focused on the language of the therapist. NLP therapists focus on the predominant sub-modalities in their clients' language. They can be visual, auditory, kinaesthetic, olfactory, or gustatory, reflecting our basic senses. For example, a client with a visual language may report seeing what you say or having a vision of the future, whereas kinaesthetic language is wrapped around touch and feelings (Bandler & Grinder, 1975).

Comparing his use of Clean Language to that of Ericksonian and NLP practitioners David Grove in Grove & Panzer (1989, p.8) states: "The shape and the structure of questions will limit how a client can respond and can leave a form of tunnel vision which will restrict his response pattern. Were we to ask: 'And how did you feel about that?' this type of question would tend to limit the client's response, and may well presuppose that the most valued way to respond pertains to a feeling, whereas the client may have wanted to return in another way (i.e. cognitively)." When Clean Language is used the client can manifest his language, and the language used by the therapist is facilitatory, allowing the client to enter into and remain within their matrix of experience.

Grove found that his clients often used personal metaphors to describe their painful emotional states and traumatic memories. He also found that when the metaphors were examined they became idiosyncratic, with meaning that only applied to his client; the metaphors had form and structure that had consistent internal logic.

Sullivan & Rees (2008, p.13) comment on Grove's work: "Rather than people having metaphors, it's as if they were their metaphors. And when these changed, they did too." This statement is a forceful assertion and, if true, could have a significant impact on the therapeutic process, which is often about change (or perhaps the hope of change). On reflection, I wonder why after thirty years there is very little research and literature on the use of metaphor in therapy; only three books have been published on Grove's work. His work has been used predominantly in the world of coaching post-2000.

Grove found that clients have many ways of describing their experiences and inner realities. They can be expressed as memories, metaphors, symbols and semantics. In Grove & Panzer (1989) he explains that the therapist should be aware of the client's predominant language. Each of the four styles is explored further:

Memories. Memory is a recall of any event of the past. Grove suggests that the client using memory language will not only relate to particular events that have occurred (real or imagined) but they may also express anticipatory memories about the future. An example is provided in Grove & Panzer (1989, p.4) "when a husband is talking about his relationship with his wife, and he focuses on 'If only I had not hit her then, we would not be having this difficulty now', the husband is identifying this past event as the main cause that is affecting him now". The client in this mode will link current issues to past events.

Symbols. Grove refers to symbols as internal and idiosyncratic rather than Jungian universal symbols. These symbols appear to be embodied within the client. Again in Grove & Panzer (1989, p.4) I find an example with the same client who faces a relationship problem with his wife: "I'm so upset with the fact that I hit my wife and my relationship now, it is so tense it is like every time we meet and talk it's like I have got this knot in my stomach." Clients expressing in this mode relate predominantly to their physiology, although "it's like a knot" is also a metaphor.

Metaphors. Grove defines metaphors in this context as individually derived, based on the client's own experience and external to their body. Remaining with the husband client, a further example is provided in Grove & Panzer (1984, p.5): "Ever since it happened we have no communication. It's like there is a wall between us. Every time I to talk to my wife, it's like trying to talk through a brick wall. The wall is the metaphor." Here the language is dense in metaphor. The issue is not feelings, memories or what has happened. He is concerned with the wall between them. According to Grove, working on the wall will promote therapeutic change and I believe will bypass any defences the husband has.

Semantics. The importance of this mode is the private definition of the words. The words carry the meaning for the client in this form of expression. For example, the husband might say "Well, she really deserved what I did to her, she is so immature. If she was more mature then we would not have a problem. The problem is really in the communication. It does not really have anything to do with what I did, but we are not communicating and I just think it is her immaturity that does it" (Groves & Panzer, 1984, p.5). Grove admits that he finds this kind of client 'tricky'; other modes are delivered with meaning whereas with semantics it is the words and not the meaning that have the most effect. This kind of client appears to be 'in their head' and perhaps not the kind of client that would grasp the concept of metaphor or symbols. Perhaps this type of client would be better suited to a more cognitive approach.

When using a predominantly metaphoric language, it is the idea of the metaphor and its visualisation (the wall) that is important. When referring to a metaphoric symbol, it is not the words that are important, it is the feelings (the knot in the stomach). In the language of memories; it is the memory itself again rather than the words that are significant.

Grove ascertains that when a client accesses their 'matrix of experience' internally through the language of metaphors, symbols and memories and the therapist utilises 'Clean Language', the client will enter into an alpha state (an altered state or conversational trance state.) It is a fascinating concept and it supports my experience of working with clients in this way. Unfortunately, no research findings are quoted to support this statement, although David Grove produced a number of videos of his client work (see Grove, 1991a/b & 1992) where it is clear that his clients expressing in metaphorical language enter a 'conversational trance' state (the Alpha State). In my experience counselling clients will often drift in and out of this state during sessions, for example when they are recalling memories, or go inward and focus on the body.

Examples of others working this way include; Deep Emotional Processing therapy (Berger, 2000) and Reichian therapy (West, 1994), psychotherapies which help clients resolve issues by following feelings and embodied experiences. Human Givens practitioners value working in this state and promote the positive changes that occur during REM/Alpha states. Griffin & Tyrrell (2013) believe that this 'right-brain' activity is a crucial element of REM sleep to complete unresolved issues and patterns through the unconscious. Metaphor work can feel like you are following a client through a dream, completing patterns, travelling from dark caves to sunny meadows.

Sullivan & Rees (2008, p.14) highlight the importance of the link between the use of Clean Language and metaphor in therapy: "As a complete approach, Clean Language can be combined with the metaphors a person uses, creating a bridge between their conscious and unconscious minds. This can become a profound personal exploration: a route to deeper understanding of themselves, and to resolution and healing". The therapist using Clean Language and metaphor can get to a new level of understanding.

In itself Clean Language is not a way of understanding metaphors; it allows the client to dwell on their experience and reveal new insights for themselves. It is not the universal answer and is inappropriate to use it in specific (more cognitive) therapies and for clients who don't readily use metaphor, symbols or memory. Its use could easily be extended to other spheres, such as education, research and commerce. Interestingly, I carried out a small piece of research on Clean Language in research interviews. The findings revealed that when the researcher asked a 'Clean' neutral question the responding answer had much more detail and revealed rich data.

Clean Language questions are only asked by the therapist that enhance the client's understanding of their experience, thus structuring an internal reality. Examples of Clean questions would be "What would you like to have happen?" or "What needs to happen?" as opposed to "What do you want?" which are aimed at the current time frame.

The neutrality of Clean Language is explained in Grove (1991b, p.9): "Clean Language is information-centred. It is neither client nor therapist centred." The use of the Clean Language, as suggested, keeps the client within their 'matrix of experience' and in a creative 'right-brain' state with the potential for the therapeutic movement of the metaphor.

Case Study ~ Clean Language and a Metaphor of Movement

It may be useful to illustrate the utilisation of Clean Language and metaphor in therapy through the introduction of a case study of a single session in my private practice. When the client accessed the internal metaphors, she drifted into a conversational trance with closed eyes and apparent Rapid Eye Movement (REM) processing. I would suggest that the use of Clean Language allowed the client to stay within the metaphor of movement.

In her forties, Emily (who is also featured in the 'Swinging Chair' case study) presented with issues around weight gain. She explained that she had received counselling and psychotherapy over many years and hinted at possible significant childhood difficulties. She indicated that she believed that her weight problems and low self-esteem were connected to childhood issues, but didn't want to go over her past with me again at this time as she had already done this with her previous therapist with little improvement. She believed that her previous therapy had been useful but hadn't got to the root of her problems. We contracted to use metaphor as a way of resolving her issues without the need to resurrect painful memories.

The session can be reviewed (almost verbatim) as follows:

Client: "When I think of my weight, it's like a bubbling black liquid" (pointing at her stomach).

Therapist: "So it's like a bubbling black liquid – and it's in your stomach?"

Client: "Yes and underneath the bubbling black liquid there is turmoil."

Therapist: "And when there is bubbling black liquid with turmoil underneath, what needs to happen?" *(I could have investigated 'turmoil' in more depth here.)*

Client: "It needs to turn white."

Therapist: "And when it needs to turn white, can it turn white?

Client: (Pause ... client has her eyes closed at this stage) "No, it's blue."

Therapist: "What needs to happen next?"

Client: "I need to get rid of some rubbish."

Therapist: "And when you need to get rid of some rubbish, what needs to happen?"

Client: "I need to put the bags in the bin."

Therapist: "How many bags do you need to put in the bin?" *(Deepening her matrix of experience.)*

Client: "Twenty-eight."

Therapist: "And when you need to put twenty-eight bags in the bin – can you put twenty-eight bags in the bin?"

Client: (Pause and Rapid Eye Movement.) "Yes" (Small sigh.)

Therapist: "And when you have put the twenty-eight bags in the bin, what happens next?"

Client: "It's still blue."

Therapist: "And when it's still blue, what needs to happen?"

Client: (Pause.) "There are seventeen more bags."

Therapist: "And when there are seventeen more bags, what needs to happen to those bags?"

Client: "They need to go in the cosmic bin."

Therapist: "And when they need to go into the cosmic bin, can they go into the cosmic bin?"

Client: (Pause and REM) "Yes." (Sigh).

Therapist: "And when they have all gone in the bin, what happens next?"

Client: "It's now light blue."

Therapist: "So what needs to happen?"

Client: "There are seventy-two layers."

Therapist: "What needs to happen to seventy-two layers?"

Client: "They need to go."

Therapist: "And when they need to go, can they go?"

Client: (Pause and REM.) "No".

Therapist: "So when there are seventy-two layers, what needs to happen?"

Client: "I need to ask for help."

Therapist: "And when you need to ask for help, who or what can help?" (Often resources can be found in this altered state – they range from 'higher-self', 'spiritual-self' or God.)

Client: "The wise part can help."

Therapist: "Thank you wise part, thank you for helping. Now what needs to happen?"

Client: "They need to go into the cosmic bin – ten at a time."

Therapist: "And when they need to go into the bin ten at a time – can they go into the cosmic bin ten at a time?"

Client: (Pause and REM.) "Yes." (Large sigh and further pause.)

Therapist: "And what happens next?"

Client: (Weeping.) "It's the core, its horrible!"

Therapist: "And when there is a core and it's horrible; what needs to happen?"

Client: "It needs to be wrapped in a metal covering."

Therapist: "And when it needs to be wrapped in a metal covering, can it be wrapped in a metal covering?"

Client: (Pause and REM.) "Yes." (Sigh.)

Therapist: "What happens next?"

Client: "It needs to go into the cosmic bin."

Therapist: "And when it needs to go into the cosmic bin, can it go into the cosmic bin?"

Client: (Pause and REM.) "Yes." (More tears.)

Therapist: "And what happens next?"

Client: (smile) "It's a white shiny thing, it's white and shiny, and there is a filter over it to keep it white and shiny."

Therapist: "And when it's white and shiny, and there is a filter over it, what else needs to happen?"

Client: "I need to put it on a pedestal."

Therapist: "And when you need to put it on a pedestal, can you put it on a pedestal?"

Client: "Yes." (Smile.)

Therapist: "What else needs to happen?"

Client: "I need to keep it."

Therapist: "And when you need to keep it, can you keep it?"

Client: "Yes, it is in my heart now." (Points to her heart.)

Therapist: "And when it is in your heart, what else needs to happen?"

Client: "Nothing".

The client opens her eyes, coming back to conscious awareness and says, "Wow".

I think it would be useful to put this session into further context at this stage:

- This Clean Language session was after several sessions of rapport building. A sound relationship is preferable

before you start this work. If Clean Language and metaphors are introduced too early, the client can sometimes find them disturbing or alien.
- Emily reported some catharsis following this session, although considerable subsequent work was carried out, approximately two years of fortnightly sessions.
- Following this and other metaphor sessions, Emily was able to talk through childhood issues in a bearable way.
- Clean Language appears to help the client stay with and develop their metaphor. There is a sense of movement here similar to Meredith's case study.
- It seems to me that the presenting issue of weight was a layer of protection for Emily and also related to her relationship with her mother. Her mother was a 'feeder' and fed Emily huge portions when she visited. We did the work on the swinging basket (as above). I can only conclude that the metaphors helped my client gain control of her emotions and body language that unconsciously projected the message 'I am an adult' to her mother.
- In another metaphor session with Emily, a potential problem with metaphor usage in therapy was highlighted. I discuss the downsides of metaphor in a subsequent chapter. In this particular session, her 'child' was put in charge of what Emily would eat (in an attempt to give her some 'adult' responsibilities). This did not work out for Emily, who reported a disastrous week of eating. Her 'child' did not want this responsibility and rebelled against this. The result was a week of a poor diet of crisps and chocolate!

Before metaphor work is completed, it is important to bring the client back to their actual age. You do not want an eight year old driving a car home, which is what happened in a group a friend and colleague was running. He had a very scared and upset client who refused to return to the group.

Grove, Metaphor and Trauma

In the research I carried out it was the indirect nature of metaphors, either verbal or embodied, which appear to be useful for some clients to revisit difficult episodes without the potential of re-traumatisation. Quotes from the participants included:

- Maddie: *"It is a very creative way of dealing with major issues that are just too big to even get out there."*
- Brian: *"Metaphor creates the containment and the safety, definitely! It's another reason why I like a metaphor. The clients have ownership of their own safety."*
- Anna: *"Metaphors can be safe containers for clients. A client found it easier to talk regarding a dinner service that he had inherited and chosen which items he might want and those he would rather let go, than talking directly about what he did not want from his abusive father."*

In a world that is increasingly and unfortunately involved in terrorism and natural disasters, I believe it is essential to review the methods of therapy that claim to help with trauma. The focus on the role of the body in trauma is gaining recognition (Levine, 1997). The work of Grove aligns itself with other somatic adjuncts to psychological trauma-focused therapies. As Peter Levine (1997, p.3) emphatically concludes: "Most trauma therapies address the mind through talk and the molecules of the mind through drugs. Both of these approaches can be of use. However, trauma is not, will not, and can never be fully healed until we address the essential role played by the body". He insists that "body sensation, rather than intense emotion, is the key to healing trauma" (p.12).

With more supportive research and training Metaphor Therapy, as outlined by Grove, could offer a viable alternative to current trauma resolution models such as EMDR and Cognitive Behavioural Trauma Focused Therapy. I have certainly experienced positive results on maturing embodied metaphors in the way David Grove proposed.

One client who had suffered a terrible abusive attack had a stabbing pain in her neck whenever she recalled the event. Using Grove's model, a metaphorical knife was found in the neck, causing the pain. Going through an elaborate process of a metaphor of movement, the blade was removed from her body and used to cut the twine that tied a bag of rubbish before both the waste and the knife were thrown into a river and drifted out to sea.

Grove claimed that metaphors are foreign objects (the knife in this case) that have been imported into the body at the moment in time just before a traumatic event; often the client has frozen this moment. This is a moment he referred to as T-1. The purpose of his metaphor work is to grow the experience up from T-1 through T, into T+1 when it's all over. The healing strategy is to provide a suitable environment in which the metaphors can be commissioned to go outside and to perform a healing function outside the body of the client. Grove would question the client's metaphor (often directly) until it confessed its strengths, matures itself and externalises, so the client no longer has to hold these feelings on the inside.

Clients can change unhelpful metaphors through the psychological process of metaphor therapy into more positive ones, or become more accepting of them. Grove found that clients could recall their traumas far more easily once the metaphor work was completed. I also find that it is more comfortable for my clients to talk about the problematic episodes in their lives after the embodied metaphors have been matured (see the 'Emily' case study above).

In summary, the use of Clean Language can be useful to help clients remain within their experience and development of a metaphor of movement. David Grove employed a remarkably soporific voice when working with clients with Clean Language. His voice was very soft and about a third of the speed of regular conversations. In my experience, this is not needed. A slightly softer tone than usual at your normal conversational pace can be useful, but not necessary.

As a therapist, you may engage fruitfully with the metaphors you create between you and your client without using Clean Language. However, in general you may find the neutralisation of your language helpful when working as a therapist. For example, the question: "How does that feel?" may throw a client because at that moment they aren't aware of feeling anything: "What is that like for you?" or "What are you experiencing?" may open up some new possibilities for you and your client.

CHAPTER THREE

METAPHORS OF ILLUMINATION

It is apparent that specific models of therapy, particularly the person-centred approaches, do not promote the deliberate and perhaps directive efforts of the therapist to influence the movement and development of a metaphor. The self-awareness that follows the illumination of the metaphor is sufficient for the client to allow them to grow and make changes for themselves.

I define metaphors of illumination as: "Metaphors that provide insight that increases one's self-awareness." I found an example of a metaphor of illumination described in Angus & Rennie (1988), in a case study in which the therapist and client worked together in the process of: "apprehending, articulating, and elaborating inner association to metaphors" (p.555). In this way, the authors of this report suggest that the counsellor appeared to help develop the client's metaphor to empathically communicate an understanding of the client's language and subjective experience.

I note that there are cultural aspects to these metaphors and they could have been misinterpreted from another cultural perspective. As Lakoff and Johnson (1980, p.233) note: "In therapy, much of self-understanding involves consciously recognising previously unconscious metaphors and how we live by them." With this mode of metaphor, it is the increased awareness of the client that is apparent, often experienced as 'light-bulb' moments.

I investigate the person-centred approach towards metaphor in more detail in a subsequent chapter. I would offer that helping a client develop their metaphor using Clean Language retains the person-centred philosophy. However, there are some clients where metaphors of relationship and illumination are sufficient. Indeed, to offer movement to someone who cannot move is at best inappropriate and at worst cruel.

The following case study of Sally provides an example of illumination and a relationship with no scope for movement. The lack of progress is a reflection of her life and where she currently finds herself.

Case Study ~ a metaphor of illumination of baths and 'Happy Days'

With some clients, it is the lack of movement that is prominent. Similar unresolved metaphors are repeated in session sometimes months or years apart. Sally was in a stressful job and an unsatisfactory marriage. In an early session, she described a dream in which she was stuck in a bath, feeling cold and exposed. Using Clean Language, I asked her what needed to happen in that dream. She needed someone to help her out; she shouted for her husband, but he was in the garage and didn't come.

Over a year later, she referred to a similar metaphor in a different guise. She said that her life was just like the Samuel Beckett play Happy Days. *She created a powerful image of herself stuck in the sand with just her head protruding. In the background was her husband, lying prostrate and lifeless on the ground. Despite some encouragement and 'Clean' questioning, Sally couldn't move the metaphor. My sense was that she was using the metaphor to tell her story of stuck-ness. Only a year after that was she able to describe a sudden change in the* Happy Days *scene. Within a fantasy, she was able to go for a brief play on the beach but had to return to her hole and stuck place.*

- This metaphor reflected Sally's life and was a metaphor of illumination for both of us. Only latterly was there a brief moment of potential change, brought about through the fantasising of moving.
- In this case, it is not the therapist's job to force a metaphor to move; this must be led by the client. It would have been tempting to help her out of the bath metaphorically or dig her out of the sand but highly inappropriate, rescuing her because of my own need to save her.
- Sally wasn't ready for the ramifications of a change of career or relationship.
- The metaphors of illumination helped her to understand the choices that she was making and the price she was prepared to pay to stay.

Therapists should be aware of the same message from the client in a slightly different metaphoric description. In my experience, some clients will keep telling their story through different metaphors until the matter is resolved or the therapist acknowledges the message and the light of illumination is recognised and accepted.

As with metaphors of relationship, metaphors of illumination can often develop into metaphors of hope or movement. There is potential for metaphors of illumination to blossom at any time, like daffodils in the springtime. Sally went on to develop her metaphor of hope. She described her relationship as being like a broken-down car which keeps going, limping along from destination to destination, often with a lot of hard work, repairs, pushing and heavy steering. The only reason the car kept going was because her children were in the back seat. When the children exit the car, she will be buying her own car and driving off into the sunset! This metaphor of hope was said with a smile on her face. She acknowledged how difficult the relationship was, why she was in this vehicle, but it would not be like this forever. Hope can be such rope to cling to during difficult times...

Dreams

Cryptophors (metaphors with a hidden meaning) and metaphors of illumination are often revealed in dreams (as evidenced in Sally's case study). The interpretation of dreams has a long history going back to Greek sleep temples. Freud (1900) believed that dreams were the conclusion-repressed wishes in the unconscious mind briefly breaking through into the conscious mind. During sleep, these wishes were typically censored by some part of the conscious mind; also during sleep, they could slip past disguised in the form of a dream.

There are many other psychodynamic theorists who contend that metaphors have a similar ability to bypass conscious censoring defence mechanisms. Freud believed these wishes were sexually based and the profound meaning of the dream could be revealed through free association, allowing the mind to roam freely on the topic of the dream. Alternatively, Carl Jung believed that symbols shown in the dream are the language of the unconscious; this is close to the widely held modern view that dreams can help the conscious mind deal with emotional problems more efficiently (Griffin & Tyrell, 2013).

Jung believed that the unconscious expressed itself through dreams. He regarded Freud's free-association technique as "misleading and inadequate use of the rich fantasies that the unconscious produces in sleep" (Jung 1964, p.11). He proposed that rather than being part of the client's defence system to deal with unwanted desires or repressed sexual feelings, dreams have an individual purposeful structure that hold an underlying intention or idea. Jung (1964, pp.43-45) recalls a dream that occurred when he was working with Freud, although he decided not reveal the contents of the dream because: "My intuition consisted of the sudden and most unexpected insight into the fact that my dream meant myself, my life, and my world, my whole reality against the theoretical structure erected by another, strange mind for reasons and purposes of its own. It was not Freud's dream, it was mine, and I understood in a flash what this dream meant."

Apart from revealing the rift between Freud and Jung and the different approaches to the interpretation of dreams, this also highlights for me the metaphorical aspect of dreams. Dreams are metaphorical in that they convey visual messages from the unconscious and contain idiosyncratic meaning and symbols. Working with dreams can be a fundamental element of a Jungian analyst's work with their client and can indicate unconscious wish fulfilment and latent transferential issues (Sharpe, 1988). Sharpe (1988, p.7) suggests that dreams indicate the individual psychical product of the individual: "The dream-life holds within itself not only the evidence of instinctual drives and mechanisms, by which those dreams are harnessed or neutralised but also the actual experiences through which we have passed ... dreams are like individual works of art."

Working with metaphors and dreams seem both to move towards bridging the gap between that which is understood and that which is beyond understanding, and there appears to be a direct link between the dream and the story that the client needs to tell (Cox & Theilgaard, 1987).

In 1953, Calvin Hall proposed that dreams are a continuation of standard cognitive processes about daily concerns. He also believed that dreams are like works of art in which ideas are transmuted into pictures. Dreams, he proposed, are a way the individual becomes more self-aware, untainted by conscious distortions and defence mechanisms. I note that this is the same process as metaphors disclosed during waking states. Viewed in this way, you could say that metaphor work is like helping your clients complete or illuminate their daydreams.

More modern research (Griffin & Tyrrell, 2014) suggests that all dreams are expressed in metaphorical imagery and that, apart from the dreamer, no one ever plays themselves in a dream; they are always just out of shot or symbolically depicted in a disguised way. During REM sleep, the unresolved concerns and undischarged emotional arousal of the previous day are discharged. You might consider that this will exclude the ongoing life concerns of the client although, as the researchers suggest, you are going to be thinking of these and becoming aroused by them most days. Griffin & Tyrrell continue to state that it is essential that dreams are metaphorical so that our memory stores don't become tainted. The emotional arousal can thus be safely discharged, the metaphorical dream can be safely forgotten, but the original record of what happened can be appropriately stored.

I contend that this is a useful way of viewing dreams and metaphors. I see dreams as metaphors. If clients mention recurring dreams, this becomes of therapeutic interest. It is apparent that the client is finding it difficult to find a metaphorical conclusion to their daily concerns through their dream state. The client can be invited to review the metaphors of the dream and contrast these to their ongoing interests (or worries of the day before the dream). This type of dream can then become a metaphor of illumination (as with Sally above) or indeed a metaphor of movement.

The concept of a co-constructed, fluid, moving metaphor is in direct contrast to the concretised metaphors which can also be used by many clients with eating disorders, who are stuck in their metaphors of 'self' (Skårderud, 2007). It also contrasts with the 'sticky metaphor' described from the psychoanalytical perspective by Mendelsohn (1989), which will keep reoccurring throughout the therapy and may result from a complementary identification in the therapist who may be experiencing similar issues or conflicts to the client.

Concretised metaphors of self (or parts of self) have been evident when working with many clients with a low sense of self. For example, with one client who came with an eating disorder, the negative internalised voice represented as a dark cave can be hard for the client to remove or change. 'Nuclear blasts' or 'beams of loving light' can fail to dent the dark cave over many sessions. Often a discussion directed at the cave can prove to be useful to illuminate its intended purpose and find an alternative role for it. The cave's purpose may be to stop the individual becoming the centre of attention or an extrovert. An alternative occupation for that metaphorical part of her may be to keep her appropriately safe. The illumination of a stuck metaphor can provide ample information for the therapist and, in turn, the client.

CHAPTER FOUR

METAPHORS OF HOPE

The absence of hope
Is like a child lost
The Agnostic Death
A fireless hearth

Life without hope
Is a life without love
The endless dark with no prospect

Hope is the chance of love
The glimpse of light
The possible
The end of hopelessness

The French social scientist Emile Durkheim explains that a society without hope would be a monstrous place incapable of promoting life. According to Durkheim, a society can no more do without this collective ideation (hope) than an organism can do without reflexes. The idea of hope is planted in human nature. As the religious and mystic poet Angelus Silesius says: "Hope is a rope, which rescues people." This metaphor refers to mystics and shaman who appear to climb a rope that appears to drop from the sky and is unattached but still holds fast. Hope defies the gravity of the everyday world and allows the individual to ascend to a higher level of reality. Thus, I propose that an essential aspect of therapy is hope.

Described by Dufault & Martocchio, (1985, p.380) as: "a multidimensional life force characterised by a confident yet uncertain expectation of achieving a future good which, to the hoping person, is realistically possible and personally significant", I would offer that from a therapeutic point of view hope is important, although to offer false hope is 'anti-therapeutic'. I would also argue that the therapist's world-view on hope and their retention of hope-full-ness spills into their work.

The training a therapist receives can also influence their beliefs around hope. To retain hope on behalf of a client during difficult times can be challenging but also very useful. If you are using up a lot of hope for your own life situation, is it more challenging to have some to spare for your clients or does this add to your empathy?

Metaphors of hope can be essential in the therapeutic process as a way in which the therapist transfers their retention of hope to their client without specifically and directly discussing the concept of hope or hopelessness. I would define them as: "Metaphors which engender a sense of hopefulness". In Cutcliffe (2004), a study that sought to explain how bereavement counsellors inspire hope in their clients, it was the implicit projection of hope and hopefulness by the therapist that was the core theme that was identified. The counsellors also suggested that the direct reference to hope was not useful.

Edey & Jevne (2003) consider that hope mostly operates as a silent factor in therapy. I propose that it is the non-literal element of metaphors that are useful and offer a sense of hope that may be useful in all types of treatment, not just bereavement work.

Jacoby (1993) suggests that it is easier to describe hope in metaphorical terms to therapy clients.

The case studies of Emily and Meredith demonstrate a sense of movement and development of the metaphors. I would suggest that this sense of action offers hope. Metaphors of landscapes and objects can change, usually to more positive ones. For example, a client who stopped smoking through use of metaphor and visualisation in therapy started their journey in a dark (unhealthy) cave and journeyed through perilous landscapes to more pleasant environments; he eventually took a shower before playing football with his children. On reflection, we both believed that the moment he stopped was when he took a shower and that the dark cave was a metaphor for his smoking or damaged lungs. The potential for movement and change from the dark to the light offers hope. If I had asked the smoker if he hoped that he could stop, he would have been doubtful initially as he had consciously tried for many years. Transferring the concretised idea of smoking to a metaphor of a cave offered the hope of change: start with the client's cave and see where he goes with it. One way of viewing this would be that existential uncertainty equates to inevitable change and change can bring hope.

I was particularly struck by the idea of engaging with metaphors and hope cognitively and somatically during my research. One participant (Yvette) concluded that therapeutic metaphors must incorporate an element of hope. For her, as a therapist who mostly used a behaviourist model, hope was at the core of the therapeutic use of metaphor. She stated that: "I never use a metaphor. I'm saying never very firmly. I can't believe that I have ever used metaphor to hinder someone. It's always about hope."

Another participant commented on his work in the palliative-care environment: "In a kind of conversational, empathy-showing way, to create therapeutic metaphors, reveals to the client that there's some kind of hope." This was so profound to hear on both occasions that I then reviewed all the metaphors disclosed in the data. Many seemed to incorporate an element of hope and, on lengthy reflection, this matched with my own experience of working with metaphor with clients. Maybe if clients can change the metaphorical representation of the 'problem' this promotes hope for them?

Yvette firmly believed that to incorporate an element of hope within her metaphors was essential. It is interesting to ask whose hope she was talking about. I think that she was talking of retaining hope in her clients by offering these metaphors, or she wanted to communicate some aspect of hope (noun) towards her client with the wish that they can retain the process of hoping (verb).

Studies have identified that the hope-oriented qualities of the therapist influence the client's level of hope (Snyder, 2000; Flesaker & Larsen, 2010). This idea fits neatly with Rogers' (1957) basic assumption that individuals, given the right conditions, tend to move towards self-healing and growth. It is apparent that sources of hope include many aspects: family; friends; spiritual beliefs; symbols, and metaphors (Hollis, Massey & Jevne, 2007; Snyder, 2004).

Meares (2005) refers to working with a deeply depressed client who had been abused in a previous relationship. In one session, she stepped out of her usual narrative and said that she wished she could be a gypsy. The therapist asked why being a gypsy would be so wonderful. The client replied that gypsies have the opportunity to travel, listen to music, dance, sing and be free. Meares (2005) proposes that this type of metaphorical musing is cathartic as it takes the individual away from their normal script and allows them an opportunity to see differently, even if for just a moment. I would add that this metaphor could have been mutually developed so that the 'normal script' might have been more meaningfully (yet indirectly) challenged.

Many different models of counselling and psychotherapy have identified that the process of hoping of the client in the process of therapy is a significant indicator of positive change (Lambert, 1992). Larsen, Edey & Lemay (2007) suggest that metaphors can be extremely potent for engendering a sense of hope in clients as an internal process or dynamic.

O'Hara (2013) advocates the use of metaphors in counselling to enable the client to recapture hope as they allow us to build a platform of new possibilities from which to challenge old dominant hopeless narratives. He suggests that the client imagine possible new ways of being through the facilitation of the 'reorganisation of self' (Lakoff, 1993) offered by metaphor utilisation. Meares (2005) compares metaphor to the external play space of the child. The adult no longer has a sandpit to play in but does have the 'mental screen of metaphor' upon which he or she can project thoughts and ideas (caves and showers) and thus retain an essential element of hope.

O'Hara (2013) suggests that many clients come to counselling because they have lost hope in relationships, self-belief, health or social functioning. Cox & Theilgaard (1987 p.45) poetically capture this sense of hopelessness: "many patients embark upon therapy because of a sense of being imprisoned within themselves for life, and fear they may never know release". Some clients expect to find hope through the process of counselling (this may also echo some form of 'standing in for metaphorically' – the therapist stands for a hope-giver). I will discuss this important aspect of what could be described as 'metaphorical transference' in a later chapter.

When hope is identified as a noun, it is made a 'thing', something to be given, taken away, transacted, and that is outside of the individual's control; for example, within a medical setting the patient links hope with cure (Eliott & Olver, 2002). This idea of hope can lead to it being regarded as an absolute, and waiting for fate to be delivered by others can result in a lack of self-efficacy (Bandura, 1982). The process of hope as a verb highlights its features, an internal process. For example, Eliott and Olver (2002) found that patients who referred to the process of hope as a verb were less dependent on the doctors' pronouncements for the maintenance of their hope; they were able to retain their hope. My friend and doctoral colleague John Pryor-Jones found in his research that hope in therapy is mostly contextual. Within a counselling service in a medical setting, for example, the narrative around hope will be more prevalent than in my private practice.

The loss of the ability to hope can be a profoundly dangerous issue. Shneidman (1985) recognises that hopelessness is the prevailing emotion in suicide. Isolation, concurrent losses and poor symptom management were listed as hope-hindering strategies in Herth (1993), a study conducted to explore strategies to foster hope with the caregivers of terminally-ill patients. This study revealed that the relational aspects of warmth, empathy and reframing were positive indicators in the hope of the client. I suggest that these are also apparent in the hope-inspiring metaphors described in many of the case studies referred to in this book, including Hannah (see below).

Case Study – Hannah's Hope

Hannah was a client in her early forties with a presenting issue of severe pain. She accessed her healing power by imagining entering a bright green cloud. Many years later she reported that she still accessed this cloud at times of distress. During one session, thinking of her "hopeless illness" situation, she started her metaphorical landscape in a deep valley and an uncomfortable path with sharp pebbles. Her journey through deep dark valleys, sharp inclines, climbing and scrambling up grey slippery cliffs, brought about many tears. She eventually found a ledge to rest on and saw a rope that was being held from the top by her husband. When she took hold of the lifeline, she was able (still with some difficulty) to climb to the top. The top was sunny and grassy; her family was waiting for her. Tears of relief followed as she realised that she was in the Hope Valley, above the village of Hope (also a real village thirty miles from my office). She had reached Hope, with a smile on her face.

Again a few comments will be useful here:

- Metaphors of light or healing colours can be cathartic for some clients in physical pain.
- The metaphor of hope was spontaneously entered into by Hannah who, like Meredith, had already enjoyed working with metaphor over many months of therapy.

- She started her metaphorical journey from an uncomfortable and painful place that matched her current situation.
- She found herself in the village of Hope, a place that is often referred to by clients. While this is an actual place close to my office, I believe that she was still within her metaphorical description/landscape. Often real locations are involved in therapeutic metaphor but I am interested in what they represent rather than their literal meaning. For example, for one client reaching the top office in a New York skyscraper was the goal of therapy. I don't believe that he wanted to move to New York; for him the office represented power and wealth.
- Hannah's husband and family were in the metaphor as she needed their help. It is often the case that outside resources are invited into the metaphor where appropriate, which also adds an element of hope (you are not on your own with this). This may not just be families; outside help has included Jesus, Buddha, spirits of passed relatives and even earth-moving machinery! The number of external resources involved can be an indication of the size of the perceived problem and also the autonomy of the client, which can be a useful indicator for the therapist. This autonomy could be regarded as positive or negative: some clients can be stubborn and 'bloody-minded' and try to resolve the metaphorical issue on their own when it appears evident that they need help; others require lots of outside support for something that could be addressed on their own. For example, decisions about their own life (e.g. careers or

relationships) needing outside support for every issue in their life.

Larsen et al. (2007) recognise that hope is a dynamic process during which counsellor and client interact in ways that can be mutually influencing; this links comfortably with the co-construction and mutual development of metaphors described earlier. Also, O'Hara (2013, p.69) contends that the deep empathic connection preferred in most therapeutic relationships (and apparent in metaphor work) is one of the critical aspects in hope generation: "When the therapist can feel with or suffer with the client, the client at some level, partly consciously, and partly unconsciously, realises that another is deeply caring for them and holding out a positive expectancy for their movement towards greater wholeness and well-being."

I would propose that counselling is a process where the client often starts with little or no hope, then introjects this as a noun from the therapist and finally holds his or her hope within as a verb. The non-literal and non-directive element of metaphors may be useful in the introjection of hope.

Metaphors of Hope and Spirituality

As with Sally, to whom I introduced you in the previous chapter, metaphors of illumination can lead to increased self-awareness; this, in turn, can help to connect that person to a spiritual connection (or re-connection). In my experience, clients who can become aware of this aspect of themselves tend to be more hopeful.

Metaphors of hope are not a new concept. The majority of religions over the centuries have offered some form of hope, forgiveness and salvation. Throughout history, it is clear that metaphors are ubiquitous in religious texts. Fran Ferder is a Franciscan sister, clinical psychologist, university professor and author. In Ferder (2010) she suggests that typologically we can relate to the central stories of the Bible involving annunciation, temptation, agony and transfiguration etc. because they are metaphorical.

Borg (2001) promotes a metaphorical reading of the bible to step away from the literal, historical and factual meanings of the text and to move beyond to the question. Borg (2001, p.41) emphasises that reading through a lens of metaphor is seeing not believing: "the point is not to believe in metaphor, but to see in the light of it ... metaphors can be profoundly true, even though they are not true. A metaphor is poetry plus, not factuality minus. That is, metaphor is not less than fact but more". He cites many biblical examples, including the story of the exodus as metaphorical narrative of the divine/human relationship depicting both the human predicament and the means of deliverance.

Rabbi David Nelson in Nelson (2006, p.20) reinforces the influence of metaphor in religion: "The value of metaphor in human life in general, and religion, in particular, can hardly be over-estimated. In our search for meaning in an overwhelming world, we use this sort of thought process to bring within our grasp that which would otherwise remain unbearably large and incomprehensible." Metaphor, according to Nelson, is a language that enables us to make sense of otherwise incomprehensible phenomena that in a spiritual dimension make sense. He also contributes to the possibility that the use of metaphor provides a subconscious awareness of what is beyond conscious understanding.

I believe that there are connections to working with clients at a deep level with metaphors that involve 'moments of deep encounter' (Tebbutt, 2014) which could be described as being 'spiritual'. For some, it is difficult to remove the spiritual from the world of therapy, as West (2010, p.1) contends: "spirituality remains at the heart of the human condition for the vast majority of humans. Even for those without such a belief, the languages and cultures of religion and spirituality retain a deep communicative expressiveness." This would appear to imply that the worlds of spirituality and metaphor sit comfortably together as expressions of humanity, otherness, non-literal communication and mutuality.

I note that when working with client's metaphors (and some quoted case studies exemplify this) there are times which equate to what Thorne (1985, p.89) describes as moments of 'tenderness' in the therapeutic relationship: "It seems as if for a space, however brief, two human beings are fully alive ... at such moments I have no hesitation in saying that my client and I are caught up in a stream of love. Within this stream there comes an effortless or intuitive understanding." Even through a secular lens, the tender moments when both the client and the therapist share in the development of a deeply personal metaphor could be described as a moment of 'relational depth', defined in Mearns & Cooper (2005, p.36) as: "A feeling of profound contact, an engagement with a client, in which one simultaneously experiences high and consistent levels of empathy and acceptance towards the Other." Perhaps the development of a unique metaphor that relates to the client's issues is a moment when high levels of empathy and acceptance are apparent?

On occasion the narrative of metaphors in the therapy room reveals spiritual aspects. For example, one woman who had recently lost both parents found her metaphor starting in a cupboard, similar to the *Tales of Narnia* story. She went through the back of the closet to a magical fairy kingdom where she could meet the spirits of her parents. During this session, she was able to say all the unsaid words she regretted not saying during their life. She found this extraordinarily cathartic and shed many tears during and after the session. Other examples include one client who met the spirits of his relatives who had died from smoking-related diseases. Their message was a clear one: stop smoking. Other clients have also reached the souls of relatives with reassuring words about their lives. For a client with health anxiety, this can elicit a considerable amount of hope.

In summary, there is ample evidence that hope is a critical component of the therapeutic process. The therapist's attitude towards hopefulness will reflect their personality and to some extent their training. Metaphors can offer a potential bridge between the therapist's hopefulness and the client's hopelessness. I posit that it is the non-literal aspect of metaphors that is helpful here. Metaphors offer a creative and non-threatening way of describing a life situation (the dark caves of our lives). Also (and crucially), the fact that the metaphors can be developed and matured into more acceptable landscapes or objects in turn offers the client an opportunity to either change or see the issue from a different angle. The option of engaging internal and external resources within the metaphor also has the potential for a more favourable outcome.

CHAPTER FIVE

METAPHORS OF RELATIONSHIP

The single most consistent view in counselling and psychotherapy is that of the importance of the therapeutic relationship. The literature on human development leads to the idea that the sense of self and self-esteem arise out of contact-in-relationship (Erskine, 2015). Relational-based therapy is underpinned by the gentle and respectful enquiry into the client's phenomenological experiences. The therapist's empathy and attunement to the client's subjective experience is vital.

Understanding how the client constructs meaning, and conveying that understanding back to the client, will inevitably include the collective awareness of each other's narrative, including metaphors. Metaphors of relationship may not involve hope, illumination or movement; they are conversational metaphors that promote a shared understanding between the parties involved in the discussion. For example, a client brings a metaphor at the beginning of our second session. She says since the first session some of the pressure in her head had reduced and left space for some positive thoughts. This feeling was unusual for her as the pressure in her head (like a pressure cooker) was always so high there was no space for positive thinking. She was reporting how it had been for her, and I gained more understanding of her. I may introduce her to this metaphor at a later stage in the therapy, but for now this is a metaphor of relationship. The client has not necessarily gained any additional self-awareness but this could have been the case if we had mutually developed the metaphor.

A dialogue in therapy implies a conversation or exchange between two (or more people). Sanders (2007, pp.111–112) encapsulates the point: "therapy is dialogue, is relational... A dialogical approach to therapy is one that emphasises or even rests completely on dialogue, that is, the co-created relationship between helper and the person being helped."

I propose a key concept in therapy is the sharing and understanding of metaphors in therapy through co-created dialogue and relationship. The empathy and improved communication when metaphors are exchanged and understood often results in the increased rapport between therapist and client (Mathieson & Hodgkins, 2005). I am defining metaphors of relationship as: "Metaphors which improve the quality of communication, empathy and rapport within a relationship". Later in this chapter I also introduce the notion of the metonymy of the therapeutic relationship. This phenomenon occurs when the client or therapist stand in for objects, concepts or others for the other party. This notion is similar but broader than the idea of transference.

From the person-centred viewpoint, metaphors in therapy appear to form part of a co-constructive dialogical process (Bohart & Tallman, 1999) that can be understood in the context of mutuality and co-creative relating (Tudor & Worrall, 2006). Tudor & Worrall (2006, p. 241–2) argue that: "Dialogue is the practice and mutuality is the outcome". Dialogue implies a conversation or exchange between two (or more people). The idea of co-created dialogue being a key concept has close connotations to the use of metaphors in therapy. "

I have argued that the use of metaphors within the therapeutic relationship, as long as they are mutually understood, helps with relational depth. The relational depth and the quality of the therapeutic relationship have long been recognised as key factors in positive outcomes for the client (Mearns & Cooper, 2005). I like a client's metaphor for describing these moments of connection, described in Knox (2011, p.132):"Do you know the ... is it, Michelangelo painting in the Sistine Chapel, where you have the two fingers? It's kind of like that, and there comes a point 'ch-ch-ch' and the contact is there..."

During my research, there were numerous references to the deepening of the relationship when metaphors are engaged with by the client and therapist. Anna was representative when she said: "I think when my clients respond to an image or a metaphor that I'm offering there is a sudden deepening of the work. It suddenly feels as if something that has been held in tension gets softened and loosened, as if the connection between the client and me feels as if it's safer to go deeper, it feels very intimate." She believed that it was the playful nature of metaphors that helped deepen the relationship: "Metaphor is playing. It is playing the way that little kids play. So maybe that's why it deepens the relationship, because it is playful. One of you is prepared to have a foot in another world. When you enter into the kid's (regressed client's) imaginative world, the adult (therapist) is the attuned care-giver who can soothe a client down."

I would add that the attuned care-giver can also share and show some understanding of the child's (client's) story (metaphor). It can be akin to jointly constructing a fairy-tale with a child when working with metaphors, a story which relates to their life situation. Not only can this illuminate and potentially offer different ways of living but the relational aspect, as alluded to by Anna, is of significant benefit.

Regression is a common occurrence in therapy and can be planned or spontaneous. Meredith spontaneously regressed in the case study described in an earlier chapter. The metaphor of a dance floor fostered the involvement of her young self. Anna talked about another client who spontaneously regressed when he likened his smoking to giving a child a lollipop. The client said: "that needy little kid who wants something and all he's getting is a sugary treat. He's not getting the real deal". In that image, he was connecting to that child within.

Maintaining contact with the regressed client is essential. Children yearn to be understood and if the client is in the psychological world of a five-year-old, the therapist plays and connects with them appropriately if attunement is to be preserved (Erskine, 2015). Brian, also a participant from my research, with a long career in Transactional Analysis summarises this topic succinctly: "One of the reasons that I like metaphor particularly is that it allows people to feel relaxed and to feel a sense of security, excitement, creativity and, most importantly, to get to parts of their unconscious or Child Ego-State. Or you might say Alpha state, so they can get to parts of themselves that they were not aware of. I think it's a very important part of psychotherapy that allows a person to take ownership of a creative part of themselves."

Case Study ~ Gravestones & Corpses

I believe that Metaphors of Relationship can be useful for all the relationships of the client. They can be particularly potent when working with couples to create a shared understanding of a phenomenon and improve communication. The Wheel of Relationship is an example of a metaphor used by me with many couples in therapy. I will often ask them to describe what would happen if they were both shipwrecked on a desert island. This reveals how their relationship works without outside influences, how collaborative they are and who takes charge etc...

In this example, I am working with Mike a young professional in his late twenties. He had a one-night stand with a work colleague and wanted to look at his relationship. I worked both with him and his partner individually and then as a couple. Throughout this process he was able to understand that his need to please others was a significant part of the issue. Also, as with all couples in therapy, the communication between them needed to improve. He avoided the potential conflict, which came with meeting his needs and his need to please everyone else, which of course wound up Michelle (his partner), particularly as she had been through a painful, unfaithful and abusive relationship before.

Several months later he was seeing me on his own again. The relationship was going well; the couple had regained some intimacy and had been on holiday. At this stage, he was unsure how often he should mention how he felt about what he had done. He didn't want to keep raking up the past but needed to accept and acknowledge what had happened. He also wanted Michelle to be able to talk about how she felt about what had happened but without living completely in the past. Feeding back his needs at this time, I found myself offering him a metaphor that came out of the blue and I had not thought of before. This metaphor was co-constructed as Mike had already given me the ingredients. "It sounds to me like you don't want to keep digging up the corpse of this event but you want to acknowledge it is there. It sounds like you need to visit the gravestone on occasion without digging up the corpse." Mike liked this analogy and wrote it down to talk to Michelle about later. He reported that the metaphor aided their communication.

On reflection, you could argue that this gave Mike a good excuse not to go back to painful events and avoid appropriate guilt. Although the gravestone appeared to offer them an appropriate place to visit and reflect on the past.

Metonymy in the Therapeutic Relationship ~ *'Standing in for'*

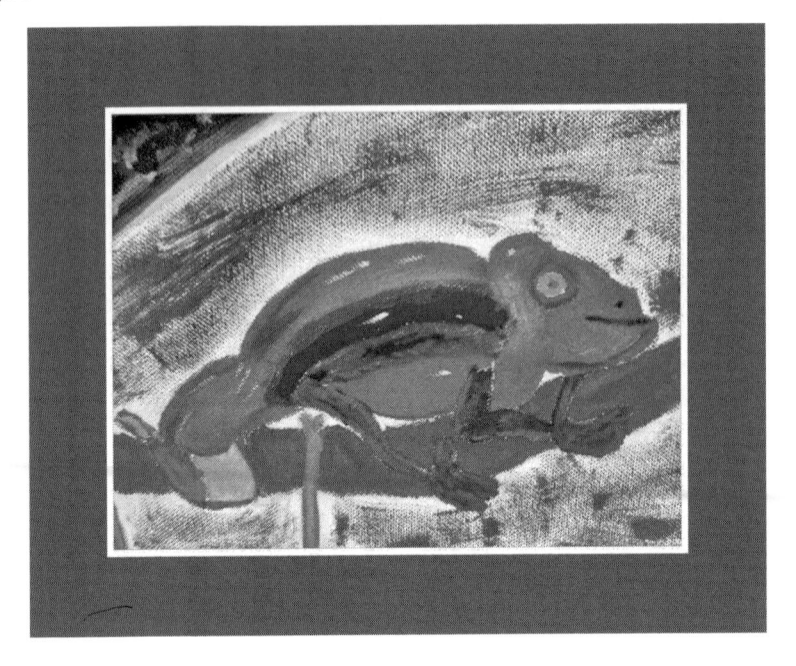

Anna, a participant in my research, made the following comment at the end of our interview back in 2012: "The therapist is a metaphor for a lover, a care-giver, a teacher, a maiden, a shaman. You are a shape shifter, a chameleon, whoever your client needs you to be." This was a 'light-bulb' moment in the research and I started an enquiry into the difference between metonym and transference in the therapeutic relationship. I was deeply moved and impacted by Anna's words as I left her interview. I reflected on the overarching concept of who therapists are metaphorically to their clients and vice versa. I remembered being a signpost to my client whose problems were the 'whale in the desert'.

This feels like an important point and relates to the 'live metaphor' concept introduced by Enckell (2002) or the Jungian notion of 'archetypal transference' (Samuels, 2003) in that the client and therapist are relating towards each other metaphorically. There have been numerous studies on transference and countertransference, and this is not the subject of this book. However, I believe it is of interest to discuss briefly in this chapter the similarities and differences of the concepts of metaphor in the sense of who the client and therapist are metonymically to each other; how they stand in for each other. I believe this is a broader concept than transference and possibly a more accessible idea for therapists who have an issue with transference. (I am principally thinking of some person-centred therapists here.)

Transference (a lynchpin of psychodynamic understanding) was originally described by Freud (1935, p.384) thus: "By this we mean a transference of feelings on to the person of the physician, because we do not believe that the situation in the treatment can account for the origin of such feelings". Transference has its roots deeply embedded exclusively in the psychoanalytical school and has metaphorical connotations. However, latterly, wider definitions have been imported by many models of therapy. Contemporary definitions of transference describe the concept as an unconscious process between a client and a therapist whereby person A acts towards person B as if person B were person C (Page & Woskett, 2015). Thus, as Jacobs (1986, p.6) alludes, transference is the term used when: "in the relationship style(s) which the client adopts towards the counsellor there are signs of past relationships".

Szajnberg (1985), from the psychoanalytical model, suggests that the German term introduced by Freud (1912) 'Ubertragung', when translated into English as 'carrying or bearing something from one place to another', can include not only the technical, critical concept embodied in the Latinate word 'transference' but also the broader idea represented in the Greek term 'metaphor'. Szajnberg (1985) posits that metaphor is the more extensive field that incorporates transference. In this paper, the author also highlights the psychological strengths of metaphor, such as its capacity for ambiguity, as an intermediate area of experience, as well as its structure, consisting of signified and signifier.

Countertransference represents the therapist's feelings, thoughts and actions used to either counter the transference projected by the client (if the client treats you like a mother figure, you find the therapist becomes masculine and business-like) or for the therapist to play out the role (in the above situation the therapist acts as the client's mother and becomes protective and angry like the client's mother did) (Hawkins & Shoet, 2006). Enckell (2002) proposes that un-mastered psychic reality reveals itself most clearly through the transference phenomenon, what he refers to as a 'living metaphor'. By focusing on the transferential living metaphor, he suggests that the therapist will learn more about the client's avoidance than by working on linguistic metaphors.

Not all therapists adopt the ideas of transference and countertransference. Indeed Shlien (1984, p.153) believes "Transference is a fiction, invented and maintained by the therapist to protect himself from the consequences of his behaviour." He proposes that the early work of Breuer & Freud relating to the first five case histories in the 1895 *Studies on Hysteria* (Breuer & Freud, 1957) involved the most sensitive older male/younger female relationships.

On the same theme, Fenichel (1941, p.95) critiques Freud (1912/1935): "Not everything is transference that is experienced by a patient in the form of affects and impulses during the course of analytic treatment. If the analyst appears to make no progress, then the patient has, in my opinion, the right to be angry, and his anger need not be a transference from childhood."

Archetypal transference is also recognised within the Jungian tradition, whereby due either to the unconscious fantasy of the client the analyst is seen as a magical healer or an evil devil, or based on the relationship the analyst and client both project unconscious material onto each other. For example, the analyst may unconsciously introduce their own 'wounded healer' into the relationship (Samuels, 2003). This is an example of a 'live metaphor' (Enckell, 2002), a metaphor apparent in the bodies of the client and therapist which is beyond language and rhetoric and is particularly apparent in the psychodynamic model.

Samuels (1985) highlights the link between Jung's alchemical imagery as a metaphor for the transference/countertransference process in psychoanalysis. He suggests that Jung's theories of interpersonal relationship can be likened to the hidden mysteries of life. Jung insisted that the psyche of the client cannot be understood in conceptual terms but only through living images or symbols, which can contain paradox and ambiguity. Alchemy reflects the process of personal transformation in the metaphor of transmuting base metals into gold. Jungians often refer to the therapist as the 'wounded healer' (Samuels, 2003) and in the

USA therapists are called 'shrinks', a derogative shorthand term for a 'head-shrinker' or shaman (Sedgwick, 2003).

Introducing an experiential viewpoint to the discussion, dependent on the stage of the therapeutic relationship, I feel it can sometimes be a brave question to ask a client "who or what am I to you?" Interestingly, Cox & Theilgaard (1987, p.19) believe that there is always an element of "risk-taking" when working with metaphors in the Aeolian mode. I have been a safe-base or safe-island to clients – and no doubt many other things if I had dared to ask. In a recent conversation that I had with a supervisee, he commented that the client reminded him of Spider Man from the first session and the image kept getting stronger – a cartoon figure that is one minute a mild-mannered studious man and the next can turn into a robust, masked individual who can trap others in his web.

I cautiously suggest that this may be different to a countertransference issue (and may have Jungian Archetypal, hero-type undertones). The client appeared to the therapist as a metaphor. The client reminded him of the cartoon figure, not necessarily a relative or friend of the client, and the therapist believed that the careful feeding back to the client of this information at the appropriate time later in the therapeutic process was of benefit.

What I am offering is that transference/countertransference (both unconscious processes which inherently introduce the symbolic) are part of a broader process in which the client and the therapist stand in for metonymically as people, objects, forms, money, gurus, wizards, signposts, parents, sisters, children, outcomes, shamans etc... For example, Rowan (1994) adds a further element to countertransference, which falls outside the traditional definition, that of 'aim attachment countertransference', where due to professional demands the therapist wants the clients to change for their own sake, not the client's. This is interesting, as this concept offers an alternative to traditional notions of transference, the symbolic 'transfer' to significant others. In other words, the client is possibly standing in metaphorically for money, status, good statistics etc. to the therapist. When Anna says: "The therapist is a metaphor for a lover, a care giver, a teacher, a maiden, a shaman. You are a shape shifter, a chameleon, whoever your client needs you to be", she is referring to a symbolic relationship in the room; there may be more than memories and emotions that connect to significant others in our past here, or what would be classically described as 'transference'.

Further research into this area may reveal interesting data into how therapists and their clients relate to and view each other metaphorically, a more palatable concept for the more person-centred counsellors who struggle with the possible avoidance of responsibility contained in the notion of transference.

I penned this poem in an attempt to capture the essence of the therapist 'standing in for'.

The therapist is a metaphor
For Lover
For Parent
For Shaman
For Chameleon
A Shape shifter
She unlocks the sacred imagination
With Nature
With Play
With Humour.

*An attuned Care-giver
A safe harbour
Contained and personal.*

*Ships, Pirates and Motorways
Frozen Gardens
Lollipops and Soup Terrines
Lead to Treasures
Healing Pathways
Bridging different worlds
Connecting rhythms and textures
Rich and deep and free.*

CHAPTER SIX

MAJOR MODELS OF THERAPY & METAPHORS

In this chapter, I review the significant models of counselling and psychotherapy and relate them to what they say (or don't say) on the subject of metaphors in therapy. The models considered include Person-Centred, Psychodynamic (including some Psychoanalytical Theory) and Cognitive Behavioural Therapy (CBT). There isn't enough space to cover all the hundreds of different models of counselling and psychotherapy so I will focus on the main players.

I suggest that the use of metaphors is apparent across all models of therapy and some models (Grove, 1989: Kopp, 1995) are based exclusively on the use of metaphor. I will comment on my categorisation where appropriate.

Person-Centred

Carl Rogers (1902-1987) was a significant contributor to the person-centred approach to therapy, which has also been adopted by many non-therapy traditions including education, architecture, environment and community building. Angus & Korman (2002, p.154) describe this model as: "a non-directive approach to therapy where the therapist's empathic understanding, genuineness and unconditional positive regard are encouraged." The core principals of this model form the basis of many therapists' work.

Rogers uses organic metaphors to describe his approaches, such as a potato that grows in a dark cellar, which reveals an organism's tendencies to self-actualise (Rogers, 1979), and seaweed on a rocky outcrop, which shows an organism's resilience in a harsh environment (Rogers, 1963). I would suggest that from a Rogerian perspective, the metaphors of therapy are mostly metaphors of illumination as they exemplify what Rogers (1967) terms the 'symbolisation of inner experiences', and the process of narrating them is a vital aspect of the person-centred approach. Some person-centred therapists would submit that illumination is sufficient within an appropriate relationship.

There are many, mostly implicit, links to the use of metaphor in the person-centred approach, as the communication of understanding is a crucial element of the model. Rogers (1973, p.4) could not be clearer when he penned: "one overriding theme in my professional life … is my caring about communication. I have wanted to understand, as profoundly as possible, the communication of the other. I have wanted to be understood." A central tenet of the person-centred model is the therapist adopting their client's "frame of reference, to perceive him, yet to perceive with acceptance and respect" (Rogers, 1953, p.41).

By its very nature, the non-directive attitude of this model lends itself to the understanding and self-development of the client, including through metaphors of illumination and relationship. The absence of the promotion of the specific use of metaphors in the person-centred literature is due to the authentic person-to-person "therapy as relationship encounter" (Rogers, 1962, p.185) stance that takes precedence over techniques and theory (Wyatt, 2001). It is about "a way of being" (Rogers, 1980, p.227), what I referred to earlier as metaphors of relationship. Although I would suggest that despite the fact that the other categories of metaphor will be engaged with by person-centred therapists, there is a minimal reference to these in the person-centred literature.

Rogers (1957) noted that clients would sometimes disagree with statements reflected to them by the counsellor, even when the statement is repeated verbatim. In this way, as Wickman et al. (1999) suggest, metaphors offer counsellors another means by which to reflect client perceptions of presenting problems.

A basic core condition of this model is empathy. I don't want to get caught up in too many definitions here, although I think it will be useful to link the empathic element of metaphors when they used appropriately in therapy. Later in this chapter, I refer to the psychodynamic model; many authors within this tradition believe that metaphor is very useful in therapy is it a way of bypassing our defence mechanisms. I talk of this later but it is interesting to note when Rogers (1975, p.3) defines empathy he also uses these words: "but not trying to uncover feelings of which the person is totally unaware, since this would be too threatening." In other words to convey empathy is essential but too much literal feedback can be threatening.

When I work with metaphors I believe that empathy can be revealed towards my client and I can enter their 'perceptual world'. I understand that metaphors can provide a means of sensing hidden feelings in a non-threatening manner. This fact may have been part of why I was drawn to this way of working in the first place.

In a study that analysed the work of Rogers from a theoretical basis, Wickman & Campbell (2003) highlight that the understanding and development of a client's metaphor can help the therapist follow the client's line of thought, improve rapport, reveal a broad sense of empathy, and can be a catalyst for change of self-perception. Cox & Theilgaard (1987), from the psychoanalytic model, also regard the use of metaphors in therapy as useful as they intensify "empathic precision" (p.6).

The literature relating to the specific use of metaphor in person-centred therapy is sparse. Exceptions include Gendlin (1978, 2003), who I believe offers a similar process to that of Grove (Grove & Panzer, 1989). However, Gendlin has been accused by certain sections of the person-centred community of being "holistic and directive" (Worsley, 2002, p.69).

Working with metaphors as a way developing a sense of mutual understanding can help the client create a truer picture of themselves in the world, while the therapist can relate to the client "in a deeply personal way" (Rogers, 1951, p.171). In this context, I would propose that the concept of 'true self' is a metaphor in itself.

Another way of considering the use of metaphor in person-centred therapy is the analysis of Rogers working with Gloria in the extraordinary session shown in a training film *The Three Approaches to Psychotherapy* in 1964 by E.L. Shostrom (Wickman et al., 1999). In this production, we have an opportunity to view Rogers at work and he appears to be co-creating and mutually developing metaphors with his client, Gloria. This, interestingly, is at odds with the person-centred literature (maybe we all end up working similarly despite the flag we operate under?).

Rogers (1953, p.34) highlights the need to fully understand the client's sense of self and to feed this knowing back to the client:

"In psychological terms, it is the counsellor's aim to perceive as sensitively and accurately as possible all of the perceptual field as it is being experienced by the client ... and having thus perceived this internal frame of reference of the other as completely as possible, to indicate to the client the extent to which he is seeing through the client's eyes."

In the filmed session Gloria, who has been recently divorced, uses language that expresses a confused sense of self; she has gone "haywire". Rogers reveals his understanding back through metaphorical language such as "no-man's land". Gloria discloses her conceptual self through the metaphor of a 'container'. Lakoff & Johnson (1980) believe that people often construct an unconscious metaphorical view of themselves as a real object, which can be a three-dimensional object such as a container. Phrases such as "I can't take any more", "I'm fit to burst", or "I'm ready to explode" are often used by people when their psychological container becomes overwhelmed. Gloria frequently refers to the parts of herself that she could accept; how she could be open to her daughter so she could see her as deep, full or whole. Gloria seems to want to work out which parts of her to keep or reject.

Rogers consistently and congruently incorporated Gloria's metaphors back into his understanding of her. He remained within her metaphorical language and framework. Throughout the session, Gloria struggles with the idea of the self that she wanted, "the perfect self". Rogers picks up on this and helps Gloria co-create a metaphor of Utopia. By reframing 'perfect', the Utopia metaphor enables Gloria to accept herself as a 'whole person' (and in some way this was a reframing of the container of parts metaphor brought by Gloria). What this seems to imply is that while metaphor isn't referred to explicitly in the person-centred literature, this film illustrates the unconscious use of metaphor by both client (to convey experiencing) and counsellor.

There appears to be a more explicit reference to the use of metaphors in the process-oriented person-centred literature. Worsley (2002) proposes that meaning is never exhausted and the client's metaphors are "radically interpersonal" (p.82). He suggests that client-generated metaphors are crucial in gaining an understanding of meaning, that they "invite shared exploration" (p.82) and the therapist needs to be guarded about what they offer into the client's metaphor. Therapist-generated metaphors are also encouraged by Worsley (2002, p.79) as a "very rich, form of asking: 'Is this what you mean?'"

Process-oriented forms of person-centred counselling, which encourage focus on the relational quality of the client's language, offer an opportunity (or permission) to work with the metaphors that are born in the therapeutic relationship. This appears to fit with much of my work and that of others, including Richard Kopp (1995). Tudor & Worrall (2006) underline the importance of metaphor to the necessary process in the understanding of experience for person-centred therapists. Rennie (1998, p.45) promotes the use of therapist-generated metaphors that: "often arise in us (therapists) when we are intensely trying to follow the path (of the client)." (I have added the bracketed words). Rennie (1998) aligns his view of the therapist-generated metaphors with that of Kopp (1995) and Gordon (1978), believing that they should fit with the client's experience and not be independent of the client. He also follows my own experience of working this way, as metaphors can be tentatively offered to clients who can reject them if the offering does not fit for them.

There is additional resonance in Rennie (1998) with other models of therapy. He suggests that the use of metaphor "liberates the secondary stream of consciousness" (p.44), which echoes the view of Freud and other psychodynamic and psychoanalytical theorists that metaphor unlocks the unconscious. (I note that it is unusual for such diverse sets of theorists to agree overtly on a topic). Rennie also raises an interesting concept that is metaphorical in itself, suggesting that the use of imagery and metaphor interrupts the horizontal flow (what I might describe as conscious conversation) of the client and takes them into a vertical descent into an aspect or memory that recaptures what was felt at the time of the trigger event.

In summary, I note that most person-centred theoretical content appears to be rooted in metaphors, for example, 'personal growth', 'journeys', 'depth of encounter' and 'true-self', although in classic Rogerian mode there is no mention of helping a client move through a metaphor. I would imagine that this would be deemed as being too directive. The person-centred therapist believes profoundly that through empathic congruent relating and illumination the client will find their way.

Psychodynamic

Here I review some of the fundamental concepts that underpin psychodynamic approaches in general, while recognising and acknowledging the diversity of approaches that fall under this heading. As with many other schools of therapy, many of the essential elements of the psychodynamic theory are metaphorical in themselves. In psychoanalytical theory, the psychic reality is thought to be ungraspable in itself; 'reality' is believed to be represented through a combination of various means, such as thought, affect sense, impressions and memories. Enckell (2002), using a psychoanalytical lens, suggests that the specific way the unconscious endeavours to represent reality is non-literal and is analogous to the theory of metaphor. Thus, a significant element of psychoanalytical investigation is comparable to the reading of metaphors.

A pivotal text to be considered here, emanating from the field of psychoanalytic theory and which I believe also has a relevance to all models of counselling and psychotherapy, is *Mutative Metaphors in Psychotherapy: The Aeolian mode* (Cox & Theilgaard, 1987). As with many psychoanalytic theories, for example, the legend of Narcissus (Freud, 1914), it takes its name from Greek mythology. The Aeolian Harp (named after the Greek god of the wind) had the capacity to pick up the 'music in the wind', a metaphor in itself for the ability to respond to the: "numerous nuances, and the hints of 'other things', which so often people offer in the therapeutic space" (Cox & Theilgaard, 1987, p.xxvi). The authors believe that the mode 'sets free' those who enjoy working creatively. It is a complex process detailed in a text of some 300 pages, although the authors provide a 'thumbnail sketch' of the essence of the model (p.xxix):

"Attend. Witness. Wait.
Discern, formulate, potentiate, and reflect mutative
metaphorical material.
Attend. Witness. Wait."

This brief description highlights the role of the therapist (and is always my intention when working with clients) as a witness to his client's unfolding experience, always waiting for possible material from their client that is rich in metaphoric potential. The Aeolian mode: "rests upon the mutative capacity (potential for movement) of metaphor and creative imagery ... and attempt(s) to make the unconscious conscious and to gradually facilitate disclosure of hidden meaning" (Cox & Theilgaard, 1987, p.96). (My brackets.)

Cryptophors, are the carriers of hidden meaning and the facilitation of disclosure of hidden meaning (illumination?), according to Cox & Theilgaard (1987), are at the centre of psychodynamic theory. The mutative potential of metaphor is the ability to perceive aspects of experience in an alternative way. Therefore, material that the client has endeavoured to avoid or deny can be brought into the client's awareness through the 'non-invasive container' of the metaphor. As Cox & Theilgaard (1987, p.99) succinctly and metaphorically describe: "Material 'filed away', appears again in the 'pending action file'."

The efficacy of the mode depends on the 'optimal synergism' between the three dynamic elements of poesies, aesthetic imperatives and points of urgency. Poesies is a process in which something is called into existence for the first time; it was not there before.

Aesthetic imperatives relate the therapist's sense (and 'imperative urge') to respond in a particular way to a client. Cox & Theilgaard (1987, p.36) explain the imperative in more detail: "The perceived coherence is one between evoked associative echoes in the therapist, the patient's clinical predicament, and the organisation of the latter's inner world." It is therefore important for the therapist to be 'fine-tuned' towards their associative resonance when working in the Aeolian mode, not to force the impulses and be mindful of non-therapeutic countertranferences that can contaminate the process.

Points of urgency indicate a perceived moment of dynamic instability, or a point of breakdown, in the patient; the patient is therefore optimally receptive to the therapist's interventions. Cox & Theilgaard (1987, p.11) indicate the optimal timing: "It is at the point of encounter when a patient becomes 'inaccessible' so that further movement seems blocked, that the mutative metaphor comes into its own". When the client is so 'heavily defended' and other interventions cannot reach them then the point of urgency, the aesthetic imperative and the dynamic of poesies (which are all inextricably related) provide the possibility of working with the Aeolian mode to unlock therapeutic possibilities inherent in metaphor and image, whereby the client is enabled to tell their story.

Cox & Theilgaard (1987) appear to be heavily influenced by the words of the Bible, Shakespeare, and other substantial authors including Dickens, Wolfe and Tolstoy. When working in this mode, the therapist seems to draw from archetypal symbols and emotions apparent in literature, myth, drama and poetry that are claimed to: "provide a powerful route of 'direct accesses' to the deepest human experience" (p.239). As he listens to his client, the therapist becomes aware of the aesthetic imperative as "archaic echoes are evoked by trigger stimuli in the patient's colloquial vernacular style of disclosure" (p.142). The resulting mutative metaphor, often informed by poetry or literature, is delivered at the perceived optimal time of the point of emergency, so that the client may safely be as close to their feelings as they can endure.

Sigmund Freud (1900) reveals that repression and censorship processes influence the unconscious to explain itself through metaphors and symbols. Freud divided the mind into the conscious mind and the unconscious mind. In his theory, the unconscious refers to the mental processes of which individuals make themselves unaware; significant psychic events take place 'below the surface'. Freud, (1917, p.295) provides us with a metaphorical description of the unconscious, conscious and the process of censorship: "Let us, therefore, compare the system of the unconscious to a large entrance hall, in which the mental impulses jostle with each other like separate individuals. Adjoining this entrance hall is a second narrower room – a kind of drawing room – in which consciousness too resides. But on the threshold between these two rooms, a watchman performs his function: he examines the different mental impulses, acts as a censor, and will not admit them into the drawing room if they displease him."

The use of metaphor in psychotherapy enhances the exchange between the unconscious and conscious realms (the entrance hall and the drawing room), as the metaphor can bypass the client's censoring defences (Cox & Theilgaard, 1987). Metaphors allow the client: "safe access to hitherto buried experience" (Cox & Theilgaard, 1987, p.69). Metaphors are the safe bridge between the conscious and unconscious and their development aids self-awareness. While Freud's notion of the unconscious and defences have been challenged by many (e.g. Sartre, 1953; Fromm, 1980; Kihlstrom, 2002), it remains at the heart of many psychodynamic models (Epstein, 1994).

Undeniably, some of Freud's concepts themselves could be regarded as being metaphorical (Draaisma, 2000). In Freud

(1900) the ego's defences stand 'on guard' to prevent any unwanted 'id' impulses from drifting from the unconscious to the conscious, a process that Freud referred to as Repression. He theorised that the ego censored objectionable items and transformed them into disguised forms, such as dreams, which are created by the unconscious through primary (thinking) processing. Desire, according to Freud (1900), our primitive driving force, is a series of metaphors that he described as displacements away from the unconscious point of origin, where one term replaces another blocked (or censored) term by one that is accepted (or uncensored) by the pre-conscious (Silverman, 1983).

Freud (1925) also found that a secondary process has a role in creating affinities and similarities that are metaphoric, although these connections are more likely to be logical, such as similes. Lichtenberg et al. (2013, p.6) highlight the importance of metaphors and imagery as forms of affective communication in which the psychoanalyst can "sense into" the internal world of their patients: "moments like a child plays with a toy."

To psychoanalysts, metaphors lie at the interchange between unconscious (primary) and conscious (secondary) ways of thinking and integrate word and image into a third entity known as "logical analogy" (Langer,1979). Metaphors are a mixture of images (primary) and words (secondary); this links with my experience that, when engaged in a metaphor, the client can be deeply connected to his unconscious processes, visualising dream-like concepts and symbols, while still being consciously aware of what he or she is experiencing.

In psychodynamic terms, metaphors are containers for powerful emotions to be processed in ways that are safe (Spandler et al.,

2013). Or, to coin a psychoanalytical term, therapeutic metaphors can be viewed as transitional objects. Winnicott (1953, p.4) describes a 'transitional object' as: "...perhaps a bundle of wool or the corner of a blanket or eiderdown, or a word or tune, or a mannerism, which becomes vitally important to the infant for use at the time of going to sleep, and is a defence against anxiety, especially anxiety of depressive type. Perhaps some soft object or type of object has been found and used by the infant, and this then becomes what I am calling a transitional object." In the same way that the infant substitutes its mother with a teddy or a thumb to reduce their anxiety through a safe indirect available container, psychoanalytically clients use metaphors to process drives and desires from censored to uncensored and acceptable concepts in a safe form.

According to Ogden (1997), analytic dialogue often takes the form of a verbal "squiggle game" (Winnicott, 1953) in which the therapist and client elaborates and modifies the metaphors that the other has unselfconsciously introduced. This 'intersubjective construction' has strong connections to the co-construction and mutual development of metaphors that I have referred to.

Carl Jung (1875-1961), the Swiss psychiatrist, introduced Analytical Psychology and coined essential concepts such as the Collective Unconscious and Jungian Archetypes. According to Young-Eisendrath & Hall (1991), Jung believed that emotionally-infused imagery is the primary organiser of the human psyche and metaphors and imagery link to the client's pre-verbal developmental stage. He also developed further the Freudian idea of the unconscious and acknowledged that specific symptoms of pain and neurosis are symbolic, e.g. sickness can relate to not being able to digest an unpleasant fact.

Jung crafts his metaphor: "As a plant produces its flower, so the psyche creates its symbols" (Jung, 1964, p.53).

I believe that there are strong links between the symbols and metaphors created by the psyche in that they create a level of meaning in a way that is represented as something else. Both employ the integration of imaginative and linguistic processes (Cox & Theilgarrd, 1987). Jung (1964, p.41) describes a symbol as a phenomenon that: "always stands for something more than its obvious and immediate meaning". He expands on their non-logical element: "Symbols, moreover, are natural and spontaneous products. No genius ever said: 'Now I'm going to invent a symbol.' No one can take a more or less rational thought, reached as a logical conclusion or by deliberate intent, and then give it 'symbolic' form... But symbols I must point out do not just occur in dreams. They appear in all kinds of psychic manifestations."

From a psychodynamic viewpoint, Robert Hobson (1985, p.65) adds to the discussion in that: "Symbols help us conceive things." Barth (1977, p.87) suggests that: "Symbol reveals the deepest mysteries of human life." I understand that there is a link between the symbols that Jung believed were messages and ideas from the unconscious (that are mostly revealed in picturesque and vivid imagery in dreams) to the deep metaphorical visualisations that clients access during therapy, referred to in Rice (1974) as evocative reflections. I find that symbols, dreams, imagery and visualisation are all metaphoric messages from the unconscious that help us conceive the world in a meaningful and safe way and connect our emotions with the visual. Metaphors are a form of symbolism (Cox & Theilgaard, 1987; Knowles & Moon, 2006).

Jung (1964, p.85) believed that as humans' contact with nature has reduced this has been compensated for by universal symbols connecting us with our "original nature". He described these symbols as archetypes, symbols with no known origin. Such symbols are also described in Young-Eisendrath & Hall (1991, pp. 1-2) as a "primary imprint (indicating) a universal disposition to construct an image, usually in an emotionally aroused state". The archetype symbols relate in the main to mythology and religion. Campbell (1986, p.55) recognised that "every myth ... whether or not by intention, is psychologically symbolic. Its narratives and images are to be read, therefore not literally, but as metaphors."

May (1991, p.38) highlights the importance of universal archetypes: "Each of us, by virtue of our pattern of myths, participates in these archetypes; they are the structure of our human existence". Kopp (1995, p.127), from the psychodynamic model, also explores this concept: "Each of us develops a personal mythology that is reflected in our personal 'guiding fiction', i.e. the metaphoric pattern that connects and makes sense of our experience of the world."

Metaphorical patterns are not only essential in explaining psychodynamic and psychoanalytical thinking, but also in the way all humans make sense of their world.

Cognitive Behavioural Therapy (CBT)

Aaron Beck developed Cognitive Therapy (CT) as a set of therapeutic techniques to help clients solve existing issues by recognising and changing dysfunctional thinking patterns. Beck noticed that when his clients adopted more adaptive ways of thinking and reacting there was a reduction in their symptoms.

CT is a form of therapy that, at its core, suggests that current negative beliefs relate to current problems such as anxiety and depression (Beck, 1979). CT has strong links with CBT, as the central theme of both models is that disordered thinking leads to maladaptive emotions and behaviours (Dobson, 2009).

Beck (1979, p.38) refers to "The Burglar example" as a therapist-generated metaphor that highlights how we can change perspectives and resultant behaviour through changing our beliefs and thoughts. In this metaphor, the client is invited to think of someone (not themselves) alone at home asleep when they suddenly hear a crash in another room. The client is invited to explore his possible thoughts and feelings and how that might lead to types of (protective) behaviour. A second version is then introduced by the therapist with the same scenario, although this time the person thinks: "The window has been left open and the wind has caused something to fall over". The aim of the metaphor is twofold, firstly to highlight that we can think differently about the same experience, and to illustrate how we think affects the way we feel and behave.

Stott et al. (2010) explain the importance of metaphors in CBT: "Cognitive Therapy has, as a central task, the aim of transforming meaning to further the client's goals and help journey towards a more helpful, realistic and adaptive view of the self and the world. Metaphor should therefore be a powerful companion" (p.14).

Whilst Stott et al. (2010) state good reasons to pay close attention to the client's own metaphors, they concede that the majority of metaphors in CBT are introduced by the therapist. Indeed, the greater part of their book *Oxford Guide to Metaphors in CBT: Building Cognitive Bridges* prescribes 'useful therapist-generated metaphors' for certain classes of psychological issues such as Eating Disorders, Psychosis, and Bipolar Disorder. For example, the metaphor of a pressure cooker is suggested as useful for those clients suffering from anger issues, as it illustrates the process of pressure building up during periods of non-assertiveness.

There are many 'empowering metaphors' suggested that relate to current scenarios in films and books that could be useful for clients. For example, Gollum's multiple internal voices heard at increased times of stress in *The Lord of the Rings* can be a helpful metaphor for those clients hearing voices. I think that these metaphors are 'issue based' rather than 'person based'. In my practice I prefer to run with my client's metaphors, although there may be occasions when I will introduce a metaphor if I feel it fits with the immediacy of the therapy. The client can always reject it, or it can open up more material if I am open to 'getting it wrong'.

Recent research into the use of football metaphors in the 'It's a Goal!' therapeutic group-work programme is revealed in Spandler et al. (2013). Using self-report questionnaires and the 'Warwick-Edinburgh Mental Well-being Scale', positive outcomes (positive affect, functioning and relationships) were reported by the 117 participants. The programme was loosely based on the CBT model and primarily aimed at men with mental-health needs, who are traditionally seen as 'hard to reach' and engage with. The research found that the use of football metaphors was seen as useful as it aided initial engagement; the language used was changed as the participants were 'players' and the facilitators were 'coaches'. The team approach also improved mutual support, and the interconnectedness within the group promoted new personal insights for the players on how they could help themselves and others, enabling self-understanding. The authors of the report believed that the metaphors provided an indirect, safe distance from which to look at themselves, motivating change.

I concur with Cox & Theilgaard (1987, p.122), who suggest that an important factor in engaging with metaphors is that the client feels "at home" in the language; although novel, the topic remains familiar. One possible interpretation of the success of this programme is that it was primarily due to the creation of an environment by the facilitator or therapist in which the members of the group were able to use their metaphors, metaphors that they could co-create, develop mutually and they could all relate to in a language that they understood. This notion is in contrast to metaphors that are generated from outside of their experience or social world, which do not resonate. It appears to me that the group mutually matured metaphors that were taken from the culture to which all the members belonged.

In a US meta-analysis study, Friedberg & Wilt (2010) report on how metaphors make CBT more accessible to young children who have limited logical reasoning skills. They suggest that this is because metaphors provide analogies for young children; they provide a simple way to understand complex reasoning techniques such as tests of evidence, reattribution, and de-catastrophising. The authors set out guidelines for good clinical practice for CBT therapists using metaphors with young clients; as with Stott et al. (2010) most are 'issue-based' rather than 'person based'. For example, 'anger volcano metaphors' are suggested for the therapist to introduce to describe the client's simmering anger. Also, 'depression is a like a bad hair day' is provided as an example to illustrate the mood's temporary and changeable quality (Blenkiron, 2005). However, Friedberg & Wilt (2010, p.104) do conclude that: "Metaphors and stories need to be individualised to match a child's individual circumstances, ethno-cultural context, and developmental level."

In summary, in CBT there appears to be a disconnection between the over-reliance on therapist-generated metaphors for specific presenting issues versus the 'person-centred' acknowledgement that: "A good metaphor reaches a child (client) where they live and fits both their internal and external reality" (Friedberg & Wilt, 2010, p.105). (I have added the bracketed word). This appears to be a technical/mechanical process in which the therapist delivers the metaphor to the 'issue' rather than the client, the human is somewhat missed.

In my experience these forms of metaphor can be more useful in group or workshop situations where generalisations are more appropriate. This formula of therapist-generated metaphor is different in energy and intention to the isomorphic metaphor aimed at the person of the client and inspired by the rhetoric within the therapy room or the client's novel metaphor.

It is apparent that the use of metaphor is cross-modular, in a similar vein to the importance of child-development and attachment theory. I suspect that the majority of therapists, irrespective of which model they follow, will have an interest in their clients' metaphors and feel guided at times to offer their own in an attempt to reveal empathy and understanding. However, how each model engages with metaphors in their writing substantially differs. In the person-centred approach it is a way of being, understanding, relationship and empathy. The psychodynamic world talks of bypassing defence mechanisms, dreams, myths and archetypes, while the CBT literature offers a plethora of therapist-generated options for different types of presenting issues.

CHAPTER SEVEN

BRINGING IT ALL TOGETHER, PRACTICAL ASPECTS

In this chapter, I focus on practical examples of how to work with metaphors in therapy. Rather than offering a linear process that should be followed strictly, I detail some guides and observations that I hope you will incorporate into your own practice in a way that works for you and your clients irrespective of your chosen model of working.

Metaphors may well up from you and your client in the moment, and this is your opportunity to hold them, play with them and develop them. You may also decide to devote some specific time to develop a particular metaphor, particularly if you are running out of time in the session. You may choose to open up a space and see what happens. For example, if a client has a specific issue or problem, you could offer them some time (perhaps using a blank canvas as described below) to see where their mind wanders on a particular topic. It will not take long for them to engage in metaphoric language, and then you are both up and running.

Let us first consider the environment you are in and how this can affect metaphors.

Environment

I was struck by how the local environment can affect the metaphors that people use when I was presenting the interim findings of this research to the University of Chester. I suddenly realised when I stood and stared at the magnificent Roman city walls outside the university that a significant number of people in the audience had talked about wall metaphors when working with their clients. Comments such as "it's like you're hiding behind a wall", or "it feels like you are breaking the wall down brick by brick" were conveyed. Maybe this was just a coincidence, although I did get to reflect on how metaphors are powerful methods of communication that have a 'to hand' quality to them. Heidegger (1962/1927) proposes that we are situated in a world of things, people and language. In this context, the audience in Chester were uniquely experiencing 'being-in-the-world' in Chester in January 2014 and impacted by the environmental and relational aspects of that experience. As Hopper (2003) found, even a painting on the therapy room wall can have a significant impact on the therapeutic relationship and process.

In my office there is an absence of family photos and academic certificates although you will find a large wooden elephant, the painting which is the front cover of this book and other items such as pebbles and pine cones, etc. The large wooden elephant, standing eighteen inches tall and two foot long, often encourages deep and meaningful conversations about the 'elephant in the room'. Perhaps what is being skilfully avoided can be gently uncovered with humorous energy.

The environment you work in is a metaphor in itself. A client of mine used to decide which cupboard in my therapy room she would open at the start of each session. One cupboard metaphorically contained all of her self-esteem issues, the other an ex-partner. We were also able to discuss how the two were connected and talked of the imaginary passage in the roof connecting the two 'walk-in' cupboards (an example of a co-constructed movement metaphor). I am proposing here that we cannot avoid the experience or the environment (the walls, elephants and cupboards).

On my placement, I worked in the filing room of a doctor's surgery. I could often see the client's file hovering behind their head on a shelf crammed with other clients' files. The size of the file related to the length of time the patient had been attached to the surgery and the number of visits they had made to the GP. I wonder if – and how – this influenced me in some way when I was talking to a 'thick file' in one session, and then a 'thin file' walked in. This no doubt relates in some way to the metonymical aspect of the relationship: the file was standing-in-for the client.

Cox & Theilgaard (1987) provide numerous examples when metaphors in group therapy were influenced by the setting. For example, noises outside led the metaphoric narrative from gravediggers to concentration camps. Snow outside can be readily linked to snowing on the inside. The participant, Maddie, who worked in a cancer-care unit, identified that none of her clients used metaphors in their narrative. She thought this was because, as cancer patients, they had been through the medical model, 'the white-coat syndrome', and they were tired of their metaphors not being heard and worked with by their doctors and nurses. There had been no co-construction of metaphors. She believed that they had given up on using metaphors as they had been misunderstood in the medical-model process. David, who also worked in a palliative care environment, only reported using his own metaphors in his interview.

This suggestion may be more of a reflection of how David and Maddie worked or recalled their memory of working.

However, there is evidence that supports Maddie's thought process of doctors and other medical-model professionals missing their patients' metaphors. In an illuminating study by Skelton et al. (2002), transcripts from 373 consultations were analysed for metaphoric content. This research revealed that in many ways doctors and patients were not speaking the same language since the metaphors they used differed. This links to the notion that some metaphors do not travel 'cross-culturally' (Lakoff & Johnson, 1980, 2003). Doctors tended to use mechanical metaphors (the urinary tract was the 'waterworks', joints suffered 'wear and tear'); they spoke of themselves as controllers of disease (they 'administered' medication, 'managed' symptoms, and 'controlled' disease) and problem-solvers (symptoms were 'clues' to be 'solved'). Patients, on the other hand, employed a wider range of vivid metaphors to describe their symptoms ('I'm like the cotton-wool man' for a sense of feeling out of touch), and metaphors of pain were used differently.

Perhaps you could be more mindful of the environment you find yourself working in. There may be improvements that could be made to improve the odds on your client offering you their metaphor or opportunities for the co-construction of metaphors.

Starting the Process

There is often a burning need within me to share a metaphorical image as a client talks. The good news with metaphors is that they are non-threatening to the client who is free to ignore, reject or amend my offering. Therapists have an obligation, I feel, to offer to work or develop the client's novel metaphor. This can be done tentatively and with Clean Language: "What is that like?" or "Tell me more about x." If you miss a client's metaphor, I find that they will introduce it to you again later, often in a different guise. It's as if their subconscious is keen to communicate metaphorical images or stories to you.

Sometimes life itself can be a metaphor for the client's predicament. For example, Sally's car really did malfunction but didn't completely break down. This was seen by the client and the therapist as a metaphor for being stuck.

I agree with other models of therapy that promote the identification and development of client-generated metaphors (Grove & Panzer, 1989; Kopp, 1990). In the structural model of existential therapy detailed in Spinelli (2005), the client's metaphor is a crucial element in the process of self-description that in itself, Spinelli posits, promotes the prospect of therapeutic change. Rather than force my metaphors on my client, I always prefer to work with those offered by the client in the therapeutic space. I believe that client's metaphors are the 'gold standard' of opportunity to commence the metaphor-maturation process. They represent the client's world-view, language, culture, dreams, fantasies and subconscious processes.

Become interested in your clients' dreams, fantasies, stories, recalled TV shows, movies, theatre and books; they all hold potential for a juicy metaphor that can illuminate or offer a positive change. Sharpe (1988, p.7) suggests that dreams indicate the individual psychical product of the individual: "dreams are like individual works of art." The client can be helped to paint their dream (or work of art) to gain some form of understanding of its metaphorical message. Occasionally the client may need to develop the work of art if the dream is incomplete. Clean Language questions can create an environment for the illumination of the dream's metaphor and to inquire if the metaphor needs to be developed. "What needs to happen?"

Blank Canvasses

The title of this book is *Whales in the Desert* and is the result of me offering a blank landscape (in this case a desert) to my client for him to complete his metaphor. In the absence of him giving me his metaphor, this is a relatively 'clean' way of getting the metaphor started. I have found that deserts can present a blank canvas, although I have also used meadows or rivers as I find that clients prefer you to use natural landscapes to complete their metaphors, even clients who spend most of their time in city centres. They will more often than not start their own metaphors in nature too. I remain unsure why this is the case; maybe we are generationally hard-wired to the natural world around us.

I asked my client to place himself in a desert and describe what was in front of him (that described his issue). This can be a useful way of accessing unconscious material and gauging how the client perceives his problems. In my experience, the size of the metaphor represents the size of their problem. This way of measuring the client's perception of their issues can be advantageous at early stages of therapy and can be an ongoing reference point throughout the therapeutic process. Clients with what they unconsciously perceive as small (or easily overcome) issues will have small barriers in front of them in the desert, like 'small walls' that can be easily traversed or 'quite clean' water in their rivers. Work is still required to help them over, under or through the wall or to clean up the river until the water is drinkable. Whales, however, are a different story! They are obviously big and can be extremely messy to go through.

I got a clue of my helping him to avoid addressing his issues when I was identified as a 'signpost'. This gave me valuable material; it was a metaphor that I held in my mind when working with this client in a multitude of subsequent sessions. It was indeed painful and messy to go through the whale but we both knew there was an 'exciting town' waiting to be explored beyond the whale. I often asked him where he was on his journey through the whale.

It is important to consider when the client is avoiding but this can be complicated, as in certain situations it may not be therapeutically appropriate to insist on clients facing their demons. I think that clients working with abuse, disorganised/anxious attachment styles or those who could be regarded as schizoid will need space and time, and may never want to go through the whale. With the 'whales in the desert' client, this process was prolonged and gentle, only going at a pace the client was happy with.

To counter this comment, in time-limited therapy delivered to clients without personality adaptations, this style of metaphor work can be completed in a handful of sessions.

Working on the Metaphor

Please refer to the metaphors of movement chapter, in particular the guidance and narrative on Clean Language.

It is important to invite the client into their world of imagination and dreams. Using neutral language and asking questions like "what happens next?" or "what needs to happen" or "what is that like" will help the client move through the metaphor without undue influence from the therapist.

While this is a simple idea, it needs some practice to master this technique. Often our natural inquisitiveness or need for feelings can obstruct the client from gaining full access to the potential of their metaphor.

- Always use the present tense, as the client is experiencing the metaphor at this moment even though it might relate to the past or the future.
- Aim your language at the age of your client in the metaphor. There can often be a regression and tone, and style needs to be altered accordingly. Don't use big grown-up words for children.
- It can be useful to review what clients have been called throughout their life. For example, I have gone from Jonnie as a child to Jon as a young adult, and I am now an older Jonathan. If I regressed in the metaphor, it would be appropriate to call my Jonnie but I might assume that I was in trouble if you called me Jonathan when I'd regressed to childhood!
- Check on the client's body language and tone, which always provide clues as to how they are feeling. If they

are getting too upset, it may be appropriate to end the session by asking them to move to a safe space.
- If you know that you are going to be working with metaphor, it is preferable to set up an imaginary 'safe place' where clients can go to at the end of each session. Securing their safe place can be a good way of introducing this work to your client and helps to build trust and rapport. Be mindful, however, that even when endeavouring to find the client's safe place they may still take an opportunity to go straight into a more gritty metaphor!

Clean Language questions about how they are experiencing the metaphor and what needs to happen can help you be guided by the client and segue towards the completion of the metaphor.

Appropriate Endings

I feel that appropriate endings are essential in specific metaphor work. Clients can be left befuddled if a metaphor of movement is left in limbo, or they can be left in a dangerously fragile state if they are left in a regressed age. There are many examples of clients feeling vulnerable for extended periods. You wouldn't leave a child alone in a park, so why leave them in a state when this is how they are experiencing reality?

The following points will help you avoid some of the potential pitfalls that I and others have inadvertently experienced and exposed our clients to. I am focussing primarily on working with metaphors of movement here.

Most of this type of narrative can be achieved by using Clean Language, although there are occasions when you might need to be more directive for the good of the client. Leave plenty of time if you can for developing metaphors. In my experience, it can take at least twenty minutes to work through a single novel metaphor using Clean Language.

Always return any regressed client to their actual age. A right way of achieving this is to introduce them to a corridor of time in which, as they return up the corridor, they get older and when they reach their real age, they become more conscious. You can also ensure that the client is grounded by asking them what they notice around them or engage in an adult conversation around work or the next appointment etc....

Sometimes clients return a little groggy from metaphor work. Clients can drift into an altered state when processing unconscious material. Some grounding exercises (as above) are essential before they drive home.

Endeavour to finish off the work in the session, or find somewhere safe to 'park' the client.

For example, if a client has let go of an issue into a metaphorical pebble but it is stuck on the riverbank, the issue/pebble will keep appearing (perhaps in a different metaphorical image). As a therapist working with metaphors of movement, I would rather see the pebble floating into the sea (the sea is usually beyond their psychological boundaries).

If the pebble refuses to move after some encouragement, then this is a metaphor of illumination. The client is now aware that this issue is difficult to let go of at this time for some reason.

As I explained earlier, clients often go on metaphorical journeys and it is the landscape that is the metaphor. They start off in challenging environments and usually, after some time, they end up in a very agreeable location in which they are happy to stay (usually a lovely meadow or such). It would not be ethical to end such a session if they were hanging from a cliff edge. Clean Language questions such as: "what happens next", "what needs to happen", "what happens after a long time" can help develop the landscape into a more agreeable one.

Keep checking if anything else needs to happen; when the answer is no and the client appears content, then the metaphor session is probably at an end.

CHAPTER EIGHT

NEGATIVES AND POTENTIAL PITFALLS

It is challenging to write about positives and negatives in therapy. In my experience (and indeed there is ample research to confirm this (e.g. Shick, 2007)), the therapist and client can often have completely different perceptions of the same session. There have been many times when I thought a productive and positive hour has taken place with a client and they thought the opposite, and vice versa. I have also thought that an interpretation was insightful and wise and the client was bound to have gained a breakthrough following a particularly sage intervention, and this 'wonderful insight' has completely passed by the client.

Most clients' feedback is that they appreciate the chance of 'offloading' onto someone who has no agenda within their lives. One of the aspects of metaphor is that there is an invitation for the other to join in with the metaphor or not. Metaphors are more subtle than a direct response, and no offence is taken by either party if they are not mutually developed.

However, having said all of this, there are times when metaphors can cause an issue. I have already mentioned that there can be an age-regression involved in some metaphors (particularly when working with the 'inner-child'). It is important to stabilise and 'grow' the client up before they leave the therapy room. Also, for some the overuse of metaphors can be too much, or even sickening. I recall an English teacher client on my placement threatening to leave the room if I mentioned one more metaphor! Maybe she had been teaching them all day or, as a beginner, I was getting carried away with using them.

It can require a lot of patience on the therapist's part when working with people with very fixed views of themselves, which equates to concretised metaphors of elements of 'self'. It can take several months before these internal 'dark caves' start to change.

From a psychoanalytical viewpoint, Siegelman (1990) identifies three pitfalls with using metaphors: overvaluing; undervaluing, and literalising. In the overestimating of metaphors, the therapist can become preoccupied with their use at the expense of other therapeutic processes (like I did with my teacher client). She suggests that the pursuit of metaphors should follow the agenda of the client, or be a collaboration between therapist and client.

Conversely, specific models of therapy (which might arguably include CBT) can focus in a 'concrete' way on changing cognition and behaviour and miss the more metaphoric or symbolic way in which the client sees the world. Siegelman (1990, p.128) states that: "our inability to see the hidden or implicit metaphors can prevent patients from enlarging the meaning of their experience". I agree with Meier & Boivin (2011, p.71) when they propose that clients "cannot be helped to reconstructively play with their narrative unless therapists themselves know how to play".

The third 'pitfall' (a metaphor in itself) identified by Siegelman (1990) is the literal interpretation of a metaphor when it is taken as a truth rather than an approximation of the reality. Thus, the tentative nature of exploring the complex and unknown in metaphoric terms is lost when we take metaphors too literally.

Alvesson (2010) also reminds us that there are some difficulties when working with metaphors. Firstly, an appealing metaphor may stand in the way of a less elegant but more appropriate description. I note that some clients much prefer a direct approach and less 'fluffy' language. Secondly, focusing on metaphors may take us away from deeper social meanings. Thirdly, oversimplification can follow metaphors. For example, a complex life becomes a simple 'journey' that does not reflect all of the nuances.

Cox & Theilgaard (1987) also have views on the potential negatives in this arena. The 'poorly timed' metaphor introduced by the therapist when the client is silent and 'creatively reflective' is regarded as ill-placed and contaminating and probably emanating from the therapist's anxieties. Further, metaphors should have an insightful therapeutic element and not involve a "string of avoidance" (p.111).

Following her research on client- and therapist-generated metaphors, Milioni (2007) points to the danger of the therapist using the client's metaphor as a 'silencing device'. In such cases, the client's world-view is closed down in favour of the therapist's interpretation. Cox & Theilgaard (1987, p.61) metaphorically describe this potentiality: "If the therapist is too predatory he may damage the hummingbird with the lasso". Milioni (2007) also found that the therapist can 'hijack' the client's metaphor, snatching it away from the client and putting a different meaning on it. In both these examples, which are loaded in metaphorical description, it appears to me that the movement and flow of the interactive process of mutual development have been halted by the therapist's ownership of the metaphor.

From a psychodynamic viewpoint, Arlow (1979) viewed metaphors as an unconscious attempt to distance oneself from powerful and potentially overwhelming affective experiences conceptually. I agree with Schlesinger (1982) who proposes that resistance is inevitable in psychoanalytical psychotherapy, and when the defence is couched in metaphorical communication it is preferable for the therapist to 'go with the resistance' and 'stay in the metaphor'.

While Freud's notion of the unconscious and defences have been challenged by many (e.g. Sartre, 1953; Fromm, 1980; Kihlstrom, 2002) it remains at the heart of many psychodynamic models (Epstein, 1994). There is some skill required from the therapist to know when the client is using metaphor as a way of avoiding matters that require discussion and when the client needs to stay in the metaphor to protect themselves. I would suggest that, if in doubt, the therapist should stay with the metaphor because it still offers an opportunity for self-awareness and movement.

There are reasons why a client may not wish to engage with metaphors, mainly if the therapist has introduced them. These factors may include low self-esteem or difficulty in visualising (Amundson, 1988; Siegelman, 1990). Rennie (1998) also suggests that metaphors can be used by clients as a way of avoiding conflict or as part of a power struggle with their counsellor. He provides an example in which the client (Audrey) introduces a metaphorical dream to her counsellor to avoid being asked about her outstanding assignment.

There may also be difficulties relating to specific mental health issues, for example the following of metaphors with those with Psychosis or Borderline Personality Disorders. Metaphors can make these clients extremely anxious as they may experience metaphors as a form of direct revelation of concrete, and often cold, reality (Cox & Theilgaard, 1987; Kopp, 1995; Enckell, 2002). I would add that some clients on the Autistic Spectrum can also struggle with the non-logical viewpoint that metaphors engage (Ramachandran, 2006). This notion is not universal and more logical metaphors, like simile, can make sense to some clients on the Autistic Spectrum (Happé, 1993). However, more recent research (McGuinty et al., 2012) suggests that specific metaphors introduced by the therapist may lead to reduced anxiety for High-functioning Autistic clients. Siegelman (1990) indicates that the more disturbed clients are, the less conscious choice of figurative language they have; thus they may find it difficult to engage with the therapist's metaphor. Spitzer (1997) also concluded that schizophrenic patients have difficulty in understanding the metaphoric meaning.

Negative aspects were also identified in my research. The participants Brian and Yvette warn that there are some possible problems and considerations with using metaphor. Brian was adamant that appropriate training is required for the therapist before they start to engage with their clients' metaphors. Yvette believed that they could be used by the therapist to rescue the client (Etherington, 2000) and avoid going to difficult places in the therapy: "am I then using metaphor in a way that someone might use a tissue?"

Brian also voiced one of my concerns: "The problem with metaphor ... it can be seen as a technique." I note that there are models of therapy, including Cognitive Behavioural Therapy, which uphold the introduction of therapist-generated metaphors for specific presenting issues (Freidberg & Wilt, 2010). I consider that this could be regarded as a technique that may not match with the client's experience; it can result in the therapist becoming preoccupied with the use of metaphor at the expense of other therapeutic or relational processes (Siegelman, 1990). For me, there is something about the person of the client that can be missed when the agenda of the therapist is paramount that can reveal a lack of congruence and empathy (Rogers, 1951, 1973). I am not against the therapist introducing metaphors 'from the shelf' although care needs to be taken to ensure that the client accepts the metaphor.

David stated: "I've always found some people where metaphor doesn't work or imagery doesn't work." This links to Amundson (1988) and Siegelman (1990), who both suggest that there are some people who struggle with metaphors due to low self-esteem or an inability to visualise. I have also found that there can be cross-cultural differences in the communication of metaphors and I need to be mindful of this when working with clients (Kövecses, 2003).

Alan referred to the potential of miscommunication when using metaphors: "Supposing I talk about theatrical stuff and sometimes we have to play roles in life and wear masks and act out different parts. That's fine as long as the client is comfortable with theatrical metaphors." He alludes to an important aspect of going with the immediacy of the session rather than mechanically planning your metaphors: "I've certainly heard of other therapists who were into what sounds like serious trouble in sessions where they introduced either metaphor or other techniques which were completely unsuited to a particular client. And then the client reacted very badly, and my view has been well, that's because the therapists plan things a bit in advance and think I'm going to do x, y or z in this session with this client, rather than going with their gut feelings. And then, because they're not going with their gut feelings, it is a complete mismatch with what the client is open to. And so, things really go badly wrong."

In summary, care needs to be given to how metaphors will be received by clients. Cultural differences and mental capacity should be considered. Some clients prefer to communicate in a more literal and direct way, and too much metaphorical language can be difficult to digest. Mutual understanding, with the potential for both client and therapist to be involved in the development of the metaphor, is the key to a positive therapeutic outcome.

CHAPTER NINE

CONCLUSION

During my research and in writing this book I have been surprised, excited, frustrated, fascinated, bored and engulfed by the process. Researching aspects of what Freud referred to as 'the impossible profession' can feel like nailing down fog at times. I hope that you, the reader, have gone through a similar process. Even being bored is an illumination.

When I was delivering a course on this topic in Leeds many years ago, a psychotherapist who had recently retired after decades of practice said: "When I come to think about it, all the major shifts that have occurred over the years with clients have come about in some way with a connection to metaphors." She may well have been pleasing me here, but she had no apparent reason to do so. It was interesting to hear her thoughts.

I believe that metaphors often pass us by because we are so used to using them every day. But when you stop and think about how delicate and beautiful they can be as a way of describing the difficult to describe, it is perhaps no surprise that they hold an innate power to illuminate or change us.

As I thumbed through my case notes in search of examples of metaphor use, I noticed some interesting points. I have not noted many; the vast majority of my records do not incorporate metaphors. It would appear that standard counselling sessions take place and notes on transference/countertransference and reflections on the process are more prevalent. I instinctively know that my client and I will use metaphors frequently, but even a therapist who is studying the use of metaphors only notes one in five sessions on their usage. On reflection, in my case notes I am just interested in metaphors of movement, hope and illumination. Unfortunately, metaphors of relationship are taken for granted.

In this book I have intentionally linked the use of metaphors to some of the important contemporary concepts in counselling. The relationship is at the heart of all that we do as therapists (and humans). To be human incorporates a deep need to connect and understand each other; metaphors are undoubtedly helpful in this endeavour for the majority of people. The client and therapist often seem to co-construct and mutually develop the metaphor in their relationship. Metaphors are a reflection of the culture and environment in which the relationship resides. I admire my fellow 'northerner' Robert Hobson's words on this phenomenon: "The meaning of metaphor is revealed within a personal and cultural context, within a society of utterances" (Hobson, 1985, p.60).

When working with metaphors, a level of empathy and understanding can ensure a level of deep encounter is created. This level of empathy was described by Cox & Theilgaard (1987, p.73) as "holistic, emotional and kinaesthetic". These moments could be described as connecting to the realms of the interpersonal, intrapersonal and transpersonal (Thorne, 1991; West, 2010; Tebbutt, 2014). When therapists are working with clients at this level, they are both developing a bespoke metaphor together. Something appears to happen that is purposeful and present, in the moment and without judgment, which is done mindfully, and the common understanding seems to evoke mutual empathy and understanding. This, in turn, seems to influence and deepen the therapeutic relationship.

I have considered a broader concept of transference, that is who the client and therapist stand for metaphorically. Not all therapists adopt the ideas of transference and countertransference (e.g. Shlien, 1984). This alternative way of viewing the therapeutic relationship offers potentially exciting insights into the dynamics involved in counselling and psychotherapy.

Lambert (1992) suggests that the retention of hope by the client is a significant indicator of a successful outcome in counselling. Cutcliffe (2004) also indicates that (particularly in bereavement counselling) the retention of hope by the therapists on behalf of their clients is of high importance, although the direct reference to hope can be counterproductive. Metaphors offer the non-direct opportunity of seeing the world differently (Cox & Theilgaard, 1987; Meares, 2005). I have provided some examples of metaphors of hope and many examples of metaphors of movement. The potential for change is linked to hope.

A new understanding of metaphors of movement, illumination, hope and relationship may help therapists' understand what type of metaphor is around in the room and assist with appropriate timing and context. Developing client's (or patient's) metaphors appears to be lacking in other areas outside the arena of therapy (Skelton et al. 2002). Those working within the medical model, for example, would theoretically build rapport and reduce anxiety and resistance if they were attuned to their patients' metaphors. Metaphors that augment aspects of hope could improve clinical outcomes and therapeutic change (Stern, 2004).

I met a woman recently who wanted to leave her long and destructive relationship. She said that it felt like it was "ten to twelve". She was alluding to the fact that it would not be long before she could end the relationship and she could 'start a new day/chapter'. She was using time as a metaphor. Sometimes when we met, the time had gone back to quarter to twelve or as she got closer it was five to twelve.

As for *Whales in the Desert,* to finish on a metaphor the clock has now struck midnight.

Glossary

The following terms are defined in the context of this book:-

Clean Language: a method of using words and forming questions in a neutral way that helps the recipient stay with their own language and metaphors.

Co-constructed Metaphors: metaphors that arise out of a shared social pool of symbols and meanings and are transacted within a co-constructed process and interactive relationship.

Cryptophor: carrier of hidden meaning.

Metaphor: the phenomenon whereby we talk, and potentially think about something in terms of something else. In this publication I use this term broadly and it incorporates similes and analogies.

Metaphors of hope: metaphors that engender a sense of hopefulness.

Metaphors of illumination: metaphors that provide insight, which results in an increase in one's self-awareness.

Metaphors of movement: metaphors which are matured and developed to such an extent that the parties involved become aware of alternative ways of viewing a particular phenomenon.

Metaphors of relationship: metaphors that improve the quality of communication, empathy and rapport within a relationship.

Metonym: If one thing can be said to 'be like' another, then it is a metaphor and 'stand for' is a metonym. Examples of British culturally-specific metonyms include 'the Crown', meaning the monarchy. An example of a metonym, perhaps with a more universal application, would be 'plastic' meaning credit cards.

References

Adler, A. (1927). Individual psychology. *The Journal of Abnormal and Social Psychology*, *22*(2), 116-122.

Alvesson, M. & Sandberg, J. (2011). Generating Research Questions Through Problematization. *Academy of Management Review*, 36(2), pp.247-271.

Angus, L., & Korman, Y. (2002). Conflict, coherence, and change in brief psychotherapy: A metaphor theme analysis. In S. R. Fussell (Ed.) *The Verbal Communication of Emotions: Inter-disciplinary perspectives*, (151–165), Mahwah, NJ: Lawrence Erlbaum Associates.

Angus, L., & Mio, J. S. (2011). At the "heart of the matter": Understanding the importance of emotion-focused metaphors in patient illness narratives. *Genre*, 44(3), 349-361.

Angus, L. E., & Rennie, D. L. (1988). Therapist participation in metaphor generation: Collaborative and noncollaborative styles. *Psychotherapy*, 25, 552–560.

Arlow, J. A. (1979). Metaphor and the psychoanalytic situation. *Psychoanalytic Quarterly*, 48, 363-383.

Bandura, A. (1982). Self-efficacy mechanism in human agency. *American Pyschologist,* 37(2), 122-147.

Bandler, R. & Grinder, J. (1975). *The Structure of Magic I: A Book About Language and Therapy*. Palo Alto, CA: Science & Behaviour Books.

Barth, J.R. (1977). *The Symbolic Imagination*. Princeton, N.J.: Princeton University Press.

Bayne, R. & Thompson, K.L. (2000). Counsellor response to clients' metaphors: an evaluation and refinement of Strong's model. *Counselling Psychology Quarterly*, 13(1), 37-49.

Beck, A. T. (1979). *Cognitive Therapy and the Emotional Disorders*. London: Penguin.

Berger, J. (2000). *Emotional fitness: Discovering our natural healing power*. Toronto: Prentice Hall.

Berger, P.L. & Luckmann, T. (1967). *The Social Construction of Reality: A Treatise in the Sociology of Knowledge*. Garden City: Anchor Books.

Bohart, A. C., & Tallman, K. (1999). *How Clients Make Therapy Work: The process of active self-healing*. Washington, D.C.: American Psychological Association.

Borg, M. (2001). *Taking the Bible seriously but not literally*. New York: Harper Collins.

Breuer, J. & Freud, S. (1957). *Studies on Hysteria*. New York: Basic Books.

Campbell, J. (1986). *The Inner Reaches of Outer Space: Metaphor as myth and as religion*. New York: Harper & Row.

Cooper, D.E. (1986). *Metaphor.* Oxford: Basil Blackwell.

Cox, M. & Theilgaard, A. (1987). *Mutative Metaphors in Psychotherapy: The Aeolian mode.* London: Tavistock.

Crotty, M. (1998). *Foundations of Social Research.* London: Sage.

Cutcliffe, J.R. (2004). The inspiration of hope in bereavement counselling. *Issues in Mental Health Nursing,* 25, 165-190.

Dalal, F. (1993). "Our historical and cultural cargo and its vicissitude in group analysis": Response to Lecture by Liesel Hearst. *Group Analysis,* 26 (4), 405-409.

Deurzen, E. van & Young, S. (2009). *Existential Perspectives on Supervision.* London: Palgrave.
Devereux, G. (1953). Cultural factors in psychoanalytic therapy. *Journal of the American Psychoanalytic Association,* 1, 629-655.

Draaisma, D. (2000). *Metaphors of Memory: A history of ideas about the mind.* Cambridge: Cambridge University Press.

Dufault, K. & Martocchio, B. (1985). Hope: Its spheres and dimensions. *Nursing Clinics of North America,* 20 (2), 379-391.

Dwairy, M. (1997). A biopsychosocial model of metaphor therapy with holistic cultures. *Clinical Psychology Review,* 17(7), 719-732.

Dwairy, M. (2009) Culture analysis and metaphor psychotherapy with Arab-Muslim clients. *Journal of Clinical Psychology*, 65(2), 199–209.

Edey, W., & Jevne, R. F. (2003). Hope, illness and counselling practice: Making hope visible. *Canadian Journal of Counselling and Psychotherapy*, 37(1), 44-51.

Eliott, J. & Olver, I. (2002). The discursive properties of 'hope': a qualitative analysis of cancer patients' speech. *Qualitative Health Research*, 12(2), 173-193.

Enckell, H. (2002). *Metaphor and Psychodynamic Functions of the Mind*. Unpublished Doctoral Thesis, University of Kuopio.

Epstein, S. (1994). Integration of the cognitive and the psychodynamic unconscious. *American Psychologist*, 49(8), 709-724.

Erskine, R. (2015). *Relational Patterns, Therapeutic Presence: Concepts and Practice in Integrative Psychotherapy*. Karnac: London.

Etherington, K. (2000). Supervising counsellors who work with survivors of childhood sexual abuse. *Counselling Psychology Quarterly*, 13(4), 377-389.

Ferder, F. (2010). *Enter the Story; Biblical Metaphors of our lives*. New York: Orbis.

Fenichel, O. (1941). *Problems with Psychoanalytic Technique*. Oxford Psychoanalytic Quarterly.

Flesaker, K. & Larsen, D. (2010). To offer hope you must have hope. *Qualitative Social Work,* 11(1), 61-79.

Freud, S. (1900). *The Interpretation of Dreams.* New York: Wiley.

Freud, S. (1912). The dynamics of transference. In G.P. Bauer (Ed.) *Essential Papers on Transference Analysis* (1994), Northvale NJ: Jason Aronson.

Freud, S. (1917). *Introductory Lectures on Psycho-Analysis.* London: Hogarth Press.

Freud, S. (1935). *A General Introduction to Psychoanalysis (Vol. 1).* New York: Liveright.

Friedberg, R. D., & Wilt, L. H. (2010). Metaphors and stories in cognitive behavioral therapy with children. *Journal of Rational-Emotive & Cognitive-Behavior Therapy,* 28(2), 100-113.

Fromm, E. (1980). *Beyond the Chains of Illusion: My encounter with Marx and Freud.* London: A&C Black.

Füredi, F. (2004). *Therapy Culture: Cultivating vulnerability in an uncertain age.* Abingdon: Psychology Press.

Geary, J. (2011). *I is An Other.* New York: Harper Collins.

Gendlin, G. (1978). *Focusing.* New York: Everest House.

Gendlin. G. (2003). *Focusing: 25th Anniversary Edition.* London: Rider.

Gonçalves, M. M., Matos, M., & Santos, A. (2009). Narrative therapy and the nature of "innovative moments" in the construction of change. *Journal of Constructivist Psychology*, 22(1), 1-23.

Gordon, D. (1978). *Therapeutic Metaphors*. Cupertino, CA: Meta Publications..

Griffin, J. & Tyrrell, I. (2013). *Human Givens: The new approach to emotional health and clear thinking. New Edition.* Chalvington: HG.

Griffin, J. & Tyrrell, I. (2014). *Why we dream: The definitive answer.* HG Publishing: East Sussex.

Grove, D. J. & Panzer, B.I. (1989). *Resolving Traumatic Memories; Metaphors and Symbols in Psychotherapy.* New York: Irvington.

Grove, D.J. (1991a). *In the presence of the past.* (Audio tape set and workbook). Eldon MO: David Grove Seminars.

Grove, D.J. (1991b). *And death shall have no dominion.* (Audio tape set and workbook). Eldon MO: David Grove Seminars.

Grove, D.J. (1992). *Reweaving a Companionable Past,* (Audio Tape set and workbook). Eldon MO: David Grove Seminars.

Happé, F. G. (1993). Communicative competence and theory of mind in autism: A test of relevance theory. *Cognition*, 48(2), 101-119

Hawkins, P. & Shoet, R. (2006). *Supervision in the Helping Professions*. Maidenhead: Open University Press.

Herth, K. (1993), Hope in the family caregiver of terminally ill people. *Journal of Advanced Nursing*, 18, 538–548.

Hobson, R.F. (1985). *Forms of Feeling.* New York: Routledge.

Hoffman, D. (1967). *Barbarous Knowledge: Myth in the Poetry of Yeats, Graves & Muir.* Oxford: Oxford University Press.

Holland, D. & Quinn, N. (Eds.) (1987). *Cultural Models in Language and Thought*. Cambridge: Cambridge University Press.

Hollis, V., Massey, K., & Jevne, R. (2007). An introduction to the intentional use of hope. *Journal of Allied Health*, 36(1), 52-56.

Hopper, E. (2003). Wounded bird: A study of the social unconscious and countertransference in group analysis. In E. Hopper (Ed.) *The Social Unconscious: Selected Papers.* London: Jessica Kingsley.

Jacobs, M. (1986). *The Presenting Past.* Milton Keynes: Open University Press.

Jacoby, R. (1993). 'The miserable hath no other medicine, but only hope': Some conceptual considerations on hope and stress. *Stress Medicine*, 9 (1), 61-69.

Jung, C.G. (1964). *Man and His Symbols.* London: Aldus Books.

Kihlstrom, J. F. (2002). No need for repression. *Trends in Cognitive Sciences*, 6(12), 502.

Kincheloe, J.K. (2001). 'Describing the Bricolage: Conceptualizing a New Rigor in Qualitative Research.' *Qualitative Inquiry*, 2001(7), 679-692.

Kirschenbaum, H.E. (2007). *The Life and Work of Carl Rogers*. Ross on Wye: PCCS Books.

Knox, R., Murphy, D., Wiggins, S. & Cooper, M. (Eds.) (2013). *Relational Depth: New perspectives and developments.* Basingstoke: Palgrave Macmillan.

Kopp, R. (1995). *Metaphor Therapy: Using Client Generated Metaphors in Psychotherapy.* New York: Brunner/Mazel.

Kövecses, Z. (2002). *Metaphor: A Practical Introduction. First Edition.* New York: Oxford University Press.

Kövecses, Z. (2003). *Metaphor and Emotion: Language, culture, and body in human feeling.* Cambridge: Cambridge University Press.

Lakoff, G. & Johnson, M. (1980). *Metaphors We Live By. First Edition.* Chicago: University of Chicago Press.
Lakoff, G. & Johnson, M. (1999). *Philosophy in the Flesh: The embodied mind and its challenge to western thought.* New York: Basic Books.

Lakoff, G. & Johnson, M. (2003). *Metaphors We Live By. Second Edition.* Chicago: University of Chicago Press.

Lambert, M. J. (1992). Psychotherapy outcome research: Implications for integrative and eclectic therapists. In J.C. Norcross & M.R. Goldfried (Eds.), *Handbook of Psychotherapy Integration, First Edition*, (94-129), New York: Basic Books.

Lamb-Shapiro, J. (2000). *The Bear Who Lost His Sleep.* Plainview, NY: Childswork/Childsplay.

Levine, P.A. (1997). *Waking the Tiger: Healing Trauma.* Berkeley, CA: North Atlantic.

Marlatt, G. A., & Fromme, K. (1987). Metaphors for addiction. *Journal of Drug Issues*, 17(1-2), 9-28.

Mathieson, L.C. & Hoskins, M.L. (2005).Metaphors of Change in the Context of Eating Disorders: Bridging Understandings with Girls' Perceptions. *Canadian Journal of Counselling,* 39(4), 260-274.

May, R. (1991). *The Cry for Myth.* New York: W.W. Norton.

McCurry, S. M., & Hayes, S. C. (1992). Clinical and experimental perspectives on metaphorical talk. *Clinical Psychology Review*, *12*(7), 763-785.

McGuinty, E., Armstrong, D., Nelson, J., & Sheeler, S. (2012). Externalising metaphors: Anxiety and high-functioning autism. *Journal of Child and Adolescent Psychiatric Nursing,* 25(1), 9-16.

McLeod, J. (2004). Social construction, narrative and psychotherapy. In L.E. Angus & J. McLeod (Eds.) *The*

Handbook of Narrative and Psychotherapy: Practice, Theory and Research, (351-366), Thousand Oaks, CA: Sage.

Mearns, D. & Cooper, M. (2005). *Working at Relational Depth in Counselling and Psychotherapy.* London: Sage.

Mearns, D., Thorne, B. & McLeod, J. (2013). *Person-Centred Counselling in Action.* London: Sage.

Meier, A., & Boivin, M. (2011). *Counselling and Therapy Techniques: Theory & Practice.* London: Sage.

Mendelsohn, R. (1989). The sticky metaphor in psychoanalytic therapy. *Psychotherapy: Theory, Research, Practice, Training, 26* (3), 380.

Milioni, D. (2007) 'Oh, Jo! You can't see that real life is not like riding a horse!': Clients' constructions of power and metaphor in therapy. *Radical Psychology,* 6(1), 123-135.

Moustakas, C. (1990). *Heuristic Research: Design, methodology, and applications.* London: Sage.

Munn, M. (2013). *Jimmy Stewart: The Truth Behind the Legend.* New York: Skyhorse.

Naziry, G., Ghassemzadeh, D. H., Katefvahid, M. & Bayanzadeh, D. S. A. (2010). The application and efficacy of metaphors in the process of cognitive-behavioral therapy for depressive patients. *Iran Journal of Psychiatry Behavioural Science.* 7(2), 24–34.

Neimeyer, R. A. (2002). The relational co-construction of selves: A postmodern perspective. *Journal of Contemporary Psychotherapy*, 32(1), 51-59.

Ogden, T. (1997). Reverie and metaphor. Some thoughts on how I work as a psychoanalyst. *The International Journal of Psycho-analysis*, 78, 719-732.

O'Hara, D. (2013). *Hope in Counselling and Psychotherapy.* London: Sage.

Page, S. & Woskett, V. (2015), *Supervising the Counsellor and Psychotherapist: A Cyclical Model.* London: Routledge.

Pitts, M. K., Smith, M. K., & Pollio, H. R. (1982). An evaluation of three different theories of metaphor production through the use of an intentional category mistake procedure. *Journal of Psycholinguistic Research*, 11(4), 347-368.

Råbu, M., Haavind, H., & Binder, P. E. (2013). We have travelled a long distance and sorted out the mess in the drawers: Metaphors for moving towards the end in psychotherapy. *Counselling and Psychotherapy Research*, 13(1), 71-80.

Ramachandran, V.S. (2006). Broken Mirrors: A Theory of Autism. *Scientific America,* 11, 62-69.
Rennie, D. (1998). *Person-Centred Counselling: An Experiential Approach.* London: Sage.

Rice, L.N. (1974). The evocative function of the therapist. In L.N. Rice and L.S. Greenberg (Eds.), *Innovations in Client-Centred Therapy,* (289-312), New York: Wiley.

Ricoeur, P. (1986). *La Metafora Viva*. Milan: Yaka Books.

Rogers, C.R. (1951). *Client Centred Therapy*. London: Constable Robinson.

Rogers, C.R. (1957). The necessary and sufficient conditions of therapeutic personality change. *Journal of Consulting Psychology*, 21(2), 95-103.

Rogers, C.R. (1962). Toward becoming a fully functioning person. In A.W. Combs (Ed.), *Perceiving, Behaving, and Becoming: A new focus on education*, (21-33), Washington D.C.: Association for Supervision and Curriculum Development, 1962 Yearbook.

Rogers, C.R. (1963). 'The actualizing tendency in relation to 'motives' and to consciousness. In M.R. Jones (Ed.), *Nebraska Symposium on Motivation*, (134-198) Lincoln, NE: University of Nebraska Press.

Rogers, C.R. (1967). *On Becoming a Person*. London: Constable.

Rogers, C.R. (1968). Some thoughts regarding the current assumptions of the behavioural sciences. In W. Coulson & C.R. Rogers (Eds.), *Man and the Science of Man*. Columbus, OH: Merrill.

Rogers, C. R. (1973). My philosophy of interpersonal relationships and how it grew. *Journal of Humanistic Psychology*, 13(2), 3-15.

Rogers, C. R. (1975). Empathic: An unappreciated way of being. *The Counseling Psychologist*, 5(2), 2-10.

Rogers, C.R. (1979). The Foundations of the Person-Centered Approach. *Education,* 100(2), 98-107.

Rogers, C.R. (1985). Toward a more human science of the person. *Journal of Humanistic Psychology*, 25(4), 7-24.

Rogers, C. R. (1980). *A Way of Being.* Boston, MA: Houghton Mifflin Harcourt.

Rogers, C. R. (1987). Rogers, Kohut, and Erickson: A personal perspective on some similarities and differences. In J. Zeig (Ed.), *The Evolution of Psychotherapy*, (179-187), Abingdon: Psychology Press.

Rowan, J. (1994). Do therapists ever cure clients? *Self & Society*, 22(5), 4-5.

Rüsch, N., Corrigan, P. W., Wassel, A., Michaels, P., Larson, J. E., Olschewski, M., & Batia, K. (2009). Self-stigma, group identification, perceived legitimacy of discrimination and mental health service use. *The British Journal of Psychiatry, 195*(6), 551-552.

Samuels, A. (1985). Symbolic dimensions of eros in transference–countertransference: Some clinical uses of Jung's alchemical metaphor. *The International Review of Psychoanalysis,* 12, 199-214.

Samuels, A. (2003). *Jung and the Post-Jungians.* London: Routledge.

Sanders, P. (2007). The 'family' of person-centred and experiential therapies. In M. Cooper, M. O'Hara, P. Schmid, & A. Bohart (Eds.) *The Handbook of Person-centred*

Psychotherapy and Counselling, (107-122), Basingstoke: Palgrave Macmillan.

Sapountzis, I. (2000). Runways, control towers and squiggles: The creation of shared moments in the treatment of children in states psychic retreat. *Journal of Infant, Child and Adolescent Psychotherapy, 1(1),* 115-133.

Sartre, J. P. (1953). *Being and Nothingness: An essay on phenomenological ontology.* New York: Washington Square Press.

Schlesinger, H. J. (1982). Resistance as process. In P.L. Wachtel (Ed.) *Resistance: Psychodynamic & Behavioral Approaches,* (25-44), New York: Plenum.

Sedgwick, D. (2003). *The Wounded Healer: Countertransference from a Jungian perspective.* London: Routledge.

Sharpe, E. F. (1988). *Dream Analysis: A Practical Handbook of Psychoanalysis.* London: Karnac Books.
Shlien, J. M. (1984). A countertheory of transference. In R.F. Levant & J.M. Shlien (Eds.) *Client-Centered Therapy and the Person-Centered Approach,* (153-181), New York: Praeger.

Shneidman, E.S. (1985). *Definition of Suicide.* New York: John Wiley and Sons.

Siegel. D. (2011). *Mindsight: Transform Your Brain with the New Science of Kindness.* London: Random House.

Siegelman, E.Y. (1990). *Metaphor and Meaning in Psychotherapy.* New York: Guilford Books.

Silverman, K. (1983). *The Subject of Semiotics*. Oxford: Oxford University Press.

Skårderud, F. (2007). Eating one's words: Part III. Mentalisation-based psychotherapy for anorexia nervosa—an outline for a treatment and training manual. *European Eating Disorders Review*, 15(5), 323-339.

Skelton J. R., Wearn A. M., & Hobbs F. D. R. (2002). A concordance-based study of metaphoric expressions used by general practitioners and patients in consultation. *The British Journal of General Practice*, 52(475), 114-118.

Snyder, C.R. (2000). *Handbook of Hope: Theory, measures, and applications*. New York: Academic Press.

Snyder, C. R. (2004). Hope and depression: A light in the darkness. *Journal of Social and Clinical Psychology, 23*(3), 347-351.

Spandler, H., Mckeown, M., Roy, A. & Hurley, M. (2013). Football metaphor and mental well-being: An evaluation of the It's a Goal! programme. *Journal of Mental Health, 22*(6), 544-554.

Spinelli, E. (2005). *The Interpreted World: An introduction to phenomenological psychology*. London: Sage.

Spitzer, M. (1997). A cognitive neuroscience view of schizophrenic thought disorder. *Schizophrenia Bulletin, 23*(1), 29-50.

Stanley-Muchow, J. Y. (1985) Metaphoric self-expression in human development. *Journal of Counselling and Development,* 64, 198-201.

Steen GJ, Reijinerse WG, Burgers C (2014) Metaphors influence reasoning? A follow-up study to Thibodeau and Boroditsky (2013). *PLoS One 9 (12): e113536.*

Stern, D.N. (2004). *The Present Moment in Psychotherapy and Everyday Life.* New York: W.W.Norton.

Stott, R., Mansell, W, Salkovskis, P., Lavender, A. & Cartwright-Hatton, S. (2010). *Oxford Guide to Metaphors in CBT: Building cognitive bridges.* New York: Oxford University Press.

Strong, T. (1989). Metaphors and client change in counselling. *International Journal for the Advancement of Counselling,* 12, 203-213.

Strong, T. (2003). Getting curious about meaning-making in counselling. *British Journal of Guidance & Counselling,* 31(3), 259-273.

Sullivan, W. & Rees, J. (2008). *Clean Language: Revealing Metaphors and Opening Minds.* Carmarthen: Crown House Publishing.

Szajnberg, N. (1985). Übertragung, metaphor, and transference in psychoanalytic psychotherapy. *International Journal of Psychoanalytic Psychotherapy,* 11, 53-75.

Tebbutt, J.N. (2014). *A comparative and theoretical study of moments of deep encounter in therapeutic and pastoral relationships*. Unpublished PhD thesis. The University of Manchester: Manchester.

Thibodeau PH, Boroditsky (2011) Metaphors we think with: the role of metaphor in reasoning. *Public Library of Science* 1 (6):e16782.

Thorne, B. (1985). *The Quality of Tenderness*. Norwich: Norwich Centre Publications.

Thorne, B. (1991). *Person-centred Counselling: Therapeutic and spiritual dimensions*. London: Whurr Publishers.

West, W. (1994). Post Reichian Therapy. In D. Jones (Ed.), *Innovative psychotherapy: A handbook* (pp. 131-145). Buckingham: Open University Press.

West, W. (Ed.). (2010). *Exploring Therapy, Spirituality and Healing*. London: Palgrave Macmillan.

White, M. & Epston, D. (1990). *Narrative Means to Therapeutic Ends*. New York: W.W. Norton.

Wickman, S. A., Daniels, M. H., White, L. J. & Fesmire, S. A. (1999). A "primer" in conceptual metaphor for counsellors. *Journal of Counselling & Development*, 77, 389–394.

Wickman, S.A. & Campbell, C. (2003). The Co-construction of Congruency: Investigating the Conceptual Metaphors of Carl Rogers and Gloria. *Counsellor Education & Supervision*, 43, 15-24.

Winnicott, D.W. (1953). Transitional Objects and Transitional Phenomena—A Study of the First Not-Me Possession, *International Journal of Psycho-Analysis,* (34), 89-97.

Wittgenstein, L. (1953). *Philosophical Investigations.* (Trans. G.E.M. Ascombe). New York: Macmillan.

Wordsworth, W., & Coleridge, S. T. (1800). *Lyrical Ballads: With Other Poems. In Two Volumes* (Vol. 1). TN Longman and O. Rees (Eds). London: Paternoster-Row.

Worsley, R. (2002). *Process Work in Person-Centred Therapy.* Basingstoke: Palgrave.

Wyatt, G. (2001). *Rogers' Therapeutic Conditions: Evolution, Theory, and Practice. (Vol. 1) Congruence.* Ross on Wye: PCCS Books.

Young-Eisendrath & Hall, J. (1991). *Jung's Self-Psychology: A constructivist perspective.* New York: The Guildford Press.

Yu, N. (1999). Spatial conceptualisation of time in Chinese. In Hiraga, Masako K., Chris Sinha and Sherman Wilcox (Eds.) *Cultural, Psychological and Typological Issues in Cognitive Linguistics: Selected papers of the bi-annual ICLA meeting in Albuquerqu*e, July 1995. viii, 338, 69-84.

Index

Aeolian mode, 13, 18, 26, 93, 103, 104, 105, 143
Blank canvas, 36, 117, 122
Clean Language, 2, 19, 33, 39, 41, 44, 45, 46, 47, 50, 51, 52, 57, 61, 63, 121, 122, 124, 125, 126, 127, 139, 157
Client-generated metaphors, 11, 12, 14, 21, 101, 121
Cognitive Behavioural Therapy, 14, 96, 111, 133
Countertransference, 90, 92, 93, 94, 135, 137, 147, 153
Countertransference, 91
Cox & Theilgaard, 13, 14, 17, 18, 26, 30, 35, 40, 67, 75, 93, 99, 103, 104, 105, 106, 107, 110, 114, 119, 130, 132, 136, 137
Culture, 6, 9, 11, 20, 21, 23, 27, 41, 114, 121, 136, 148
Dreams, 65, 66, 67, 68, 107, 110, 116, 121, 122, 124
Eating disorders, 13, 14, 68
EMDR, 59
Empathy, 8, 36, 38, 71, 73, 76, 81, 83, 84, 98, 99, 115, 133, 136, 140
Endings, 125
Environment, 9, 15, 60, 73, 96, 97, 114, 117, 118, 119, 120, 122, 136
Erskine, 46, 83, 86, 144
Freud, 14, 65, 66, 90, 91, 92, 102, 103, 106, 107, 108, 131, 135, 142, 145
Gendlin, 46, 47, 99, 145, 146
Griffin, 2, 51, 66, 67, 146
Grove, 2, 17, 19, 33, 45, 46, 47, 48, 49, 50, 52, 58, 59, 60, 61, 96, 99, 121, 146
Hobson, 26, 29, 35, 110, 136, 147
Humour, 43, 44
Jacobs, 1, 91, 147
Jung, 65, 66, 92, 109, 110, 148, 154, 158

Kopp, 12, 17, 21, 96, 101, 111, 121, 132, 148
Lakoff & Johnson, 4, 6, 20, 100, 120
Language games, 26
Memories, 48, 49, 50, 53, 94, 103
Metaphors of hope, 8, 71, 79, 139
Metaphors of illumination, 8, 139
Metaphors of movement, 8, 29, 139
Metaphors of relationship, 9, 83, 140
Metonymy, 84
Person-centred, 18, 35, 45, 62, 63, 84, 90, 94, 96, 97, 99, 101, 102, 115, 154
Points of urgency, 105
Psychodynamic, 2, 26, 35, 65, 90, 92, 98, 102, 103, 104, 107, 108, 110, 111, 116, 131, 144
Regression, 41, 124, 128
Relational depth, 80, 85
Ricoeur, 11, 152
Rogers,, 45, 97, 98, 99, 133, 152, 153
Semantics, 48, 50
Spirituality, 80
Symbols, 23, 48, 49, 50, 51, 65, 66, 73, 93, 106, 108, 109, 110, 139
Therapist-generated metaphors, 11, 12, 19, 20, 21, 22, 102, 112, 115, 130, 133
Tinnitus, 35, 36, 38
Transference, 75, 84, 89, 90, 91, 92, 94, 95, 135, 137, 145, 153, 154, 157
Trauma, 2, 45, 59

Printed in Great Britain
by Amazon